SpringerBriefs in Ser

Series Editor

Robin Qiu, Division of Engineering & Information Science, Pennsylvania State University, Malvern, PA, USA

Editorial Board Members

Saif Benjaafar, Industrial and Systems Engineering, University of Minnesota, Minneapolis, MN, USA

Brenda Dietrich, Cornell University, New York, USA

Zhongsheng Hua, Zhejiang University, Hefei, Anhui, China

Zhibin Jiang, Management Science, Shanghai Jiao Tong University, Shanghai, China

Kwang-Jae Kim, Pohang University of Science and Technology, London, UK

Lefei Li, Department of Industrial Engineering, Tsinghua University, Haidian, Beijing, China

Kelly Lyons, Faculty of Information, University of Toronto, Toronto, ON, Canada

Paul Maglio, School of Engineering, University of California, Merced, Merced, CA, USA

Jürg Meierhofer, Zurich University of Applied Sciences, Winterthur, Bern, Switzerland

Paul Messinger, Alberta School of Business, University of Alberta, Edmonton, Canada

Stefan Nickel, Karlsruhe Institute of Technology, Karlsruhe, Baden-Württemberg, Germany

James C. Spohrer, IBM University Programs World-Wide, IBM Almaden Research Center, San Jose, CA, USA

Jochen Wirtz, NUS Business School, National University of Singapore, Singapore, Singapore

SpringerBriefs present concise summaries of cutting-edge research and practical applications across a wide spectrum of fields. Featuring compact volumes of 50 to 125 pages, the series covers a range of content from professional to academic.

Typical publications can be:

A timely report of state-of-the art methods

A bridge between new research results, as published in journal articles

A snapshot of a hot or emerging topic

An in-depth case study

A presentation of core concepts that students must understand in order to make independent contributions

SpringerBriefs are characterized by fast, global electronic dissemination, standard publishing contracts, standardized manuscript preparation and formatting guidelines, and expedited production schedules.

The rapidly growing fields of Big Data, AI and Machine Learning, together with emerging analytic theories and technologies, have allowed us to gain comprehensive insights into both social and transactional interactions in service value co-creation processes. The series SpringerBriefs in Service Science is devoted to publications that offer new perspectives on service research by showcasing service transformations across various sectors of the digital economy. The research findings presented will help service organizations address their service challenges in tomorrow's service-oriented economy.

Qinghai Miao • Fei-Yue Wang

Artificial Intelligence for Science (AI4S)

Frontiers and Perspectives Based on Parallel Intelligence

Qinghai Miao
University of Chinese Academy of Sciences
Beijing, China

Fei-Yue Wang
Institute of Automation
Chinese Academy of Sciences
Beijing, China

ISSN 2731-3743   ISSN 2731-3751   (electronic)
SpringerBriefs in Service Science
ISBN 978-3-031-67418-1   ISBN 978-3-031-67419-8   (eBook)
https://doi.org/10.1007/978-3-031-67419-8

© The Editor(s) (if applicable) and The Author(s), under exclusive license to Springer Nature Switzerland AG 2024

This work is subject to copyright. All rights are solely and exclusively licensed by the Publisher, whether the whole or part of the material is concerned, specifically the rights of translation, reprinting, reuse of illustrations, recitation, broadcasting, reproduction on microfilms or in any other physical way, and transmission or information storage and retrieval, electronic adaptation, computer software, or by similar or dissimilar methodology now known or hereafter developed.

The use of general descriptive names, registered names, trademarks, service marks, etc. in this publication does not imply, even in the absence of a specific statement, that such names are exempt from the relevant protective laws and regulations and therefore free for general use.

The publisher, the authors and the editors are safe to assume that the advice and information in this book are believed to be true and accurate at the date of publication. Neither the publisher nor the authors or the editors give a warranty, expressed or implied, with respect to the material contained herein or for any errors or omissions that may have been made. The publisher remains neutral with regard to jurisdictional claims in published maps and institutional affiliations.

This Springer imprint is published by the registered company Springer Nature Switzerland AG
The registered company address is: Gewerbestrasse 11, 6330 Cham, Switzerland

If disposing of this product, please recycle the paper.

Preface

The landscape of scientific research has undergone a profound transformation recently, largely due to the emergence and convergence of Intelligent Science and Technology such as Artificial Intelligence (AI), deep learning, and large language modes (LLMs). Today, the movement of AI for Sciences (AI4S) is gearing up to revolutionize traditional practices of research and development through "NewSci with New IT": New Paradigms for Scientific Discoveries and Applications through Decentralized Autonomous Science (DeSci) with Intelligent Technology beyond AlphaGo, ChatGPT, or Sora likes. Clearly, AI4S represents a fast integration and pivotal injunction of natural wisdom, technological intelligence, and social smartness, and we must do our best to ensure its processes and consequences would benefit our humanity, serve our people well and better, that is, must be of the people, by the people, and for the people.

To this end, this brief book prescribes a new technological path for AI4S from the perspective of parallel science by parallel intelligence based on cyber physical social systems (CPSS), aiming at making AI4S "6S": Safe in the physical world, Secure in the cyberspace, Sustainable for the ecological development, Sensitive to privacy, individual rights, and resource utilization, Service to All, and Smartness of All. Major highlights include:

Parallel Intelligence of "Hanoi" in Three Worlds

We would like to rid people of tedious labors for knowledge generation and application in the physical world by creating a new workspace in the cyberspace of humanity for supervision and governance of knowledge distribution and automation. For this we need a world model larger than foundation models or LLMs, and base our large model, called Parallel Worlds, on Karl Popper's Three Worlds of reality and knowledge, i.e., world 1 the physical world, developed mainly by "Old IT" the industrial technology, world 2 the mental world, developed mainly with "Past IT" the information technology, and world 3 the artificial world, to be developed mainly through "New IT" the intelligent technology. In short, Parallel Intelligence is "Three Worlds, Three ITs" in CPSS that integrates Human, Artificial, Natural, Organizational, and Imaginative Intelligence into One Intelligence, the Hanoi Intelligence, where all models of intelligence would operate in parallel and in harmony.

Parallel Science by Parallel Scientists

Parallel Science is aiming at "Three Worlds, Three Sciences" for "Three Worlds, Three Knowledge," i.e., Descriptive Knowledge mainly for World 1, Predictive Knowledge mainly for World 2, and Prescriptive Knowledge mainly for World 3, by "Three Worlds, Three Scientists," i.e., Digital Scientists, over 80%, Robotic Scientists, under 15%, and Biological Scientists, less than 5%, working in parallel as integrated Parallel Scientists for knowledge discovery, knowledge generation, and knowledge automation in manned or unmanned fashion through the autonomous and self-evolving fusion of natural philosophy, social studies, and intelligent science.

A New "Day" for Parallel Research in Parallel Universes

We will conduct parallel education and parallel research in three worlds in parallel in the future. Parallel scientists would start their workdays in the morning with a new "AM" of more than 20 hours, i.e., Autonomous Mode where digital scientists and robotic scientists would complete the task of research and development under the supervision of biological scientists, move to the afternoon with a new "PM" of fewer than 3 hours, i.e., Parallel Mode where robotic scientists and digital scientists should accomplish the job under the remote support from biological scientists, and enter into the night with a new "EM" of less than 1 hour, i.e., Expert or Emergency Mode where biological scientists must be presented in fields to carry out the mission with the help of digital and robotic scientists.

    This is our vision for a future of AI4S, and call it the Science of SCE+: Slow, Casual, Enjoy, Easy, Elegant, …

    We hope you enjoy reading this booklet.

Beijing, China	Qinghai Miao
March 6, 2024	Fei-Yue Wang

Contents

1 **AI4S Based on Parallel Intelligence** 1
   1.1 Introduction 1
   1.2 Parallel Intelligence 2
      1.2.1 The Origin of Parallel Intelligence 2
      1.2.2 Parallel Systems and ACP Approach 3
      1.2.3 Parallel Intelligence 4
   1.3 Parallel Science: AI4S Based on Parallel Intelligence 5
      1.3.1 Intelligent Technology: The New IT 5
      1.3.2 Digital Scientists and Robotic Scientists 7
      1.3.3 A New "Day" for Parallel Research in Three Modes 8
      1.3.4 Cornerstones for AI4S in Industry 5.0 8
   1.4 HANOI-AI4S: A General Framework for Parallel Science 9
      1.4.1 Challenges in AI4S 9
      1.4.2 The Necessity of a General AI4S Framework 12
      1.4.3 HANOI-AI4S: A General Framework for AI4S 14
      1.4.4 AI Methods Used in AI4S 16
   References 18

2 **AI for Mathematics** 21
   2.1 Guiding Mathematical Intuition with AI 21
      2.1.1 Knowledge: Topology 22
      2.1.2 AI: Neural Networks 22
      2.1.3 Human Role: Mathematician in the Loop 23
   2.2 FunSearch: Searching New Programs for Combinatorial
       Optimization 23
      2.2.1 Extremal Combinatorics and Combinatorial Optimization ... 25
      2.2.2 AI: Genetic Programming and LLM 25
      2.2.3 Human Role: Guiders in Prompt Engineering 27
   2.3 AlphaGeometry: Proving Theorems for Euclidean Plane
       Geometry 28
      2.3.1 IMO Euclidean Geometry 29

		2.3.2	AI: Symbolic Deduction and Language Model	30
		2.3.3	Synthetic Dataset	32
		2.3.4	Human Role	33
	2.4	AlphaTensor: Discovering Faster Matrix Multiplication Algorithm		33
		2.4.1	Knowledge: Matrix Multiplication	34
		2.4.2	AI: Deep Reinforcement Learning	35
	2.5	AlphaDev: Discovering Faster Sorting Algorithms		36
		2.5.1	Design and Optimization of Algorithms	36
		2.5.2	AI: AlphaZero Paradigm	37
		2.5.3	Data: Generating Data via Self-playing	38
	References			39
3	**AI for Physics**			**41**
	3.1	OmniFold: Unfolding Observables from Large Hadron Collider		41
		3.1.1	Knowledge: Particle Physics and LHC	42
		3.1.2	AI: Parallel Intelligence, Neural Networks, and Neural Resampling	43
	3.2	Magnetic Controlling of Tokamak Plasmas		45
		3.2.1	Knowledge: Plasma Physics	46
		3.2.2	AI: Reinforcement Learning	47
		3.2.3	Human Role and Collaboration	48
		3.2.4	Summary	49
	3.3	Finding Evidence for Intrinsic Charm Quarks		49
		3.3.1	Knowledge: Parton Distribution Functions	50
		3.3.2	AI: Neural Network and Genetic Algorithms	50
		3.3.3	Collecting Experimental Dataset	51
		3.3.4	Human Role: The NNPDF Collaboration	51
	References			52
4	**AI for Biology**			**53**
	4.1	AlphaFold: Predicting 3D Protein Structure		53
		4.1.1	Domain Knowledge	54
		4.1.2	AI: Transformer and Attention Mechanisms	54
		4.1.3	Dataset	55
		4.1.4	Human Role	55
		4.1.5	AlphaFold2 and New Advancements	57
	4.2	RFdiffusion: De Novo Design of Protein		57
		4.2.1	De Novo Design of Protein Structure	58
		4.2.2	AI: Diffusion Model	59
		4.2.3	Summary	60
	4.3	scGPT: A Foundation Model for Single-Cell Biology Research		60
		4.3.1	Knowledge: Single-Cell Multi-omics	61
		4.3.2	AI: Transformer, LLM, and Foundation Models	62
	References			63

5 AI for Health and Medicine 65
5.1 Swarm Learning: Decentralized and Confidential Diagnosis 65
    5.1.1 AI: Distributed Learning and Blockchain 66
    5.1.2 Related Works on AI for Clinical Diagnostics 67
5.2 Geneformer: Predicting Candidate Therapeutic Targets 68
    5.2.1 Knowledge: Gene Regulatory Networks 68
    5.2.2 AI: Transformer for Gene 68
    5.2.3 Data: Genecorpus-30M 69
    5.2.4 PandaOmics: A Platform for Therapeutic Target
         Identification 69
5.3 drugAI: De Novo Drug Design Using Transformer with MCTS 70
    5.3.1 SMILES: The Language for Drug Discovery 71
    5.3.2 AI: Transformer and MCTS 71
    5.3.3 Graph Neural Network in AI Pharmacy 72
References 72

6 AI for Chemistry 75
6.1 AlphaFlow: Discovering Materials via Bayesian Active Learning ... 75
    6.1.1 AI: RL-Guided Self-driving Lab 76
    6.1.2 Knowledge: Colloidal Atomic Layer Deposition (cALD) 77
    6.1.3 Related Works 77
6.2 Coscientist: Autonomous Chemical Research with Large
    Language Models 78
    6.2.1 AI: Generative Pretrained Transformer 79
    6.2.2 More Related Works Based on LLMs 80
References 80

7 AI for Material Science 81
7.1 CAMEO: Discovering Materials via Bayesian Active Learning 81
    7.1.1 Knowledge: Structural Phase Map 82
    7.1.2 AI: Active Learning 82
    7.1.3 Human Role 84
7.2 GNoME: Graph Networks for Materials Exploration 84
    7.2.1 Density Functional Theory 85
    7.2.2 AI: Graph Neural Network 86
    7.2.3 Dataset 87
7.3 A-Lab: An Autonomous Laboratory 88
    7.3.1 Solid-State Synthesis of Inorganic Powders 88
    7.3.2 AI: Active Learning with Robotics 90
    7.3.3 Datasets 91
    7.3.4 Human Role 91
References 91

8 AI for Astronomy 93
8.1 Locating Hidden Exoplanets 93
    8.1.1 ALMA Observations 94

		8.1.2	Simulation for Synthetic Data	94
		8.1.3	EfficientNet and RegNet	95
	8.2	DLPosterior: Estimating Dark Matter Distribution		96
		8.2.1	Mass-Mapping and Gravitational Lensing Effect	97
		8.2.2	AI: Score-Based Generative Modeling	98
	8.3	Astroconformer: Analyzing Stellar Light Curves		100
		8.3.1	Stellar Light Curves	101
		8.3.2	AI: Self-attention and Transformer	102
	References			103
9	**Toward a Sustainable AI4S Ecosystem**			**105**
	9.1	A Brief Summary of AI4S from Viewpoint of HANOI		105
	9.2	Toward AI4S Ecosystem		107
		9.2.1	The Origins and Goals of AI4S	107
		9.2.2	DeSci	109
		9.2.3	DeSci and DAO	110
		9.2.4	DeSci with Blockchain	110
	9.3	Foundation Intelligence Based on TRUE DAO		111
		9.3.1	Federated Ecosystem	111
		9.3.2	True DAO to Intelligent Federation	112
	References			113

Chapter 1
AI4S Based on Parallel Intelligence

1.1 Introduction

As Artificial Intelligence (AI) rapidly advances in recent years, its application in various scientific fields such as physics, chemistry, biology, astronomy, and others is also emerging. This technology, known as AI for Sciences (AI4S), is revolutionizing scientific research by providing new tools and techniques for analyzing data, discovering patterns, and making predictions, etc. AI4S helps scientists process large amounts of data more efficiently, identify new research directions, and accelerate the pace of discovery. It also has the potential to address complex scientific challenges that may be difficult or time-consuming for humans to solve alone.

Although AI has found application across a wide array of disciplines, this book can only delve into a fraction of the recent advancements. Specifically, this book will cover new developments in mathematics, physics, biology, health and medicine, chemistry, materials science, and astronomy in the following chapters from 2 to 8.

In this chapter, we first provide an overview of paradigm shifts of scientific research from a new perspective of Parallel Intelligence (PI) [1–3]. We give a brief description of Parallel Science featured with three ITs for three worlds in three modes. We then discuss current challenges in AI4S from multiple aspects. To address these challenges, we propose a general framework, called HANOI-AI4S, for analyzing and guiding the development of AI4S. In this book, HANOI-AI4S serves as a uniform framework for evaluation and comparison of AI4S across different fields.

1.2 Parallel Intelligence

Currently, the AI4S paradigm is advancing swiftly, with its role differing significantly across various disciplines. Therefore, it is necessary to provide a unified framework for the review and analysis of AI4S methods in various fields. To this end, Parallel Intelligence (PI) is a suitable choice. In this section, we give a brief introduction of Parallel Intelligence, including its origins, principles, techniques, and frameworks.

1.2.1 The Origin of Parallel Intelligence

Although the idea of Parallel Intelligence (PI) can be traced back to the study of Circular Causality in 1940s, more recent research originated from the Cyber–physical–social systems (CPSS) [4] by Fei-Yue Wang.

The development of CPSS is driven by the desire to create more efficient, adaptive, and user-centric systems that can address the challenges of an increasingly interconnected and complicated world [1, 5]. Briefly, CPSS are integrated systems that combine computation, networking, and physical processes with human interactions. These systems are characterized by their ability to sense, compute, communicate, and actuate in the physical world, often with significant involvement of social elements such as human behavior, preferences, and interactions.

Figure 1.1a illustrates fundamental framework of CPSS. The philosophical foundation of CPSS is Karl Popper's "Three Worlds" theory, which posits that the universe is composed of three unified and coherent worlds: the physical world (World 1), the mental world (World 2), and the artificial world (World 3):

⋆ World 1 encompasses objective matter and phenomena.
⋆ World 2 is the subjective world of knowledge, containing human consciousness and experiences.
⋆ World 3 is the objective world of knowledge, involving the products of cultural, civilized, scientific, technological, or theoretical systems recorded and stored by various carriers.

These three worlds interact and influence each other, mapping onto the Physical Space and Cyberspace, thus forming five rings as shown in Fig. 1.1a. This framework can effectively integrate various resources and values in three-dimensional space, facilitating both "emergence" and "convergence" in complex systems [6]. Utilizing technologies like virtual reality, it integrates various physical, computational, and human brain resources existing in the physical world, the mental world, and the artificial world, realizing a management and service model that is parallel, transparent, intelligent, and ubiquitous.

1.2 Parallel Intelligence

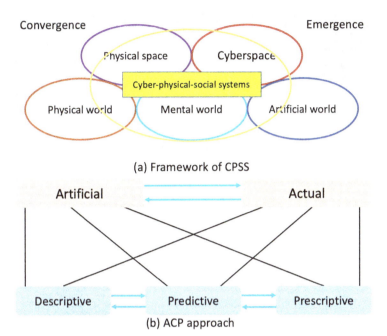

Fig. 1.1 The "five rings, five boxes" [6] for CPSS and parallel intelligence by Prof. Fei-Yue Wang

1.2.2 Parallel Systems and ACP Approach

In 2004, Fei-Yue Wang proposed the theory of Parallel Systems [7], transforming models from system analyzers to data generators, making CPSS computable, testable, and verifiable, providing a possibility to bridge the modeling gap between physical world and artificial world. In fact, as early as 1994, Professor Wang had proposed the Shadow System [8], which views models as data generators and visualization tools. Subsequently, influenced by the theory of "Three Worlds," it developed into concepts of Parallel Systems. The core of the Parallel Systems theory is the ACP approach [1], which comprises three components, i.e., artificial societies (A), computational experiments (C), and parallel execution (P). Among them, the artificial systems or societies are the foundation, computational experiments are the core, and parallel execution is the goal.

As shown in Fig. 1.1b, the relationship between artificial systems and actual systems can be one-to-one, one-to-many, many-to-one, or many-to-many, depending on the complexity of the problem and the accuracy of the solution. In the process of solving the problem, there is virtual–real interaction and interactive execution between the artificial system and the actual system, resulting in a form of intelligence called Parallel Intelligence.

The ACP approach integrates descriptive intelligence, predictive intelligence, and prescriptive intelligence as the Foundation Intelligence. The descriptive intelligence assists in constructing artificial systems, predictive intelligence facilitates computational experiments, and prescriptive intelligence offers mechanisms to guide and optimize parallel execution. Therefore, Parallel Systems can use one or more virtual (artificial) spaces to resolve the essential contradiction between complexity and intelligence, making "unsolvable" problems "solvable," and thus achieving the effective solution of the CPSS.

1.2.3 Parallel Intelligence

Knowledge plays an essential role in both AI and CPSS. Knowledge automation tries to realize the cyclic process of knowledge generation, acquisition, application, and recreation. The goal is to transform the characteristics of uncertainty, diversity, and complexity (UDC) of complex systems into the characteristics of agility, focus, and convergence (AFC) of intelligent systems [9]. To achieve this, it is essential to embed knowledge automation into the Parallel Intelligence framework and process based on ACP approach.

According to different applications, the actual system and its corresponding artificial systems in Parallel Intelligence can be connected in different modes (Fig. 1.2). By comparing and analyzing the behavior of reality and simulation, learning and predicting the future behavior of the system, and modifying the corresponding control strategies, this framework has three operation modes respectively: learning and training, experimentation and evaluation, and control and management.

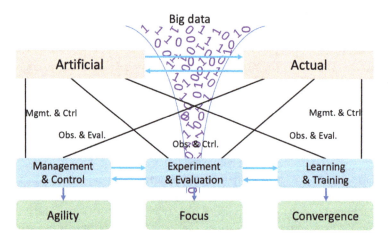

Fig. 1.2 ACP-based parallel intelligence: small data, big data, and deep intelligence

In the **learning and training mode**, artificial systems are connected or combined with actual scenarios or cases, serving as the "center" for operators and managers to learn and train. It is worth noting that the artificial system does not have to be identical to the actual system; it is a possible alternative form of evolution of the actual system in different directions.

In the **experimentation and evaluation mode**, the artificial system serves as a platform for computational experiments to analyze and predict the behavior of the actual system in different scenarios.

In the **management and control mode**, the artificial system is connected to the actual system in real-time online and replicates the actual behavior with high fidelity. By identifying operational parameters based on the differences in behavior between the actual system and the artificial system, feedback control is generated.

In conclusion, the ACP approach uses small data to produce big data and then extracts deep intelligence from big data, effectively overcoming the limitations of traditional methods and solving the contradiction between the "emergence" of phenomena and the "convergence" of solutions in complex systems.

In the framework of Parallel Intelligence, the artificial system (A) is a broad knowledge model that can be seen as an extension of traditional mathematical or analytical modeling. Computational experiments (C) provide a way to analyze, predict, and select complex decisions, representing a sublimation of simulation. Parallel execution (C) is a new feedback control mechanism composed of virtual–real interactions, guiding actions and locking targets. The closed-loop feedback, virtual–real interaction, and parallel execution between artificial systems and actual systems can effectively control complex systems, leading to the emergence of parallel intelligence [2, 3].

Accordingly, the workflow of parallel intelligence framework mainly consists of the following three steps: (1) Construct an artificial system corresponding to the actual complex system; (2) Use computational experiments to train, predict, and evaluate complex systems; (3) Achieve parallel control and management of complex systems by setting up interactions and mutual learning between the actual physical system and the virtual artificial system. Through interaction between the virtual and the real, parallel intelligence can continuously bring the actual system closer to the artificial system, simplifying the UDC challenges faced by the study of complex systems and achieving AFC management and control of complex systems, empowering the entire process of CPSS.

1.3 Parallel Science: AI4S Based on Parallel Intelligence

1.3.1 *Intelligent Technology: The New IT*

As we are seeing, AI is greatly enhancing or even redefining the productivity of industrial technology (old IT) and information technology (past IT). In other words,

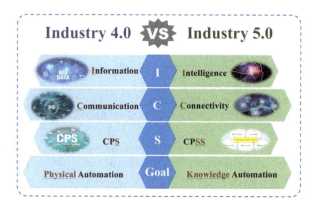

Fig. 1.3 Paradigm shifts from Industry 4.0–5.0 driven by New IT [11]

"intelligent industry" is the upgrading of existing industry with the enhancement of Intelligent Technology (new IT), thereby changing our society and promoting shifts in scientific research paradigms.

The current AI technologies and applications have clearly shown that the "intelligent industry" has begun: Big data has become the new means of production, blockchain and smart contracts have formed new production relations, and large models and robots have become new productive forces.

After "Industry 4.0," the European Union released the "Industry 5.0" document in 2021, hoping to use artificial intelligence technology to lead Europe toward a sustainable, people-oriented, and CPSS-like industrial society, as proposed in the original definition of Industry 5.0 in 2014 by Prof. Fei-Yue Wang [10]. At present, the consensus among international academic and industrial circles is that the core concepts of Industry 5.0 [6, 11] are "knowledge automation" based on CPSS and driven by intelligence, as shown in Fig. 1.3. Its essence lies in parallel technology and industries, fostering virtual and real collaboration in parallel. Its significant manifestations are seen in the forms of "new liberal arts," "new science," and "new engineering."

The main purpose of AI4S is to use "new IT" to promote changes in traditional scientific research. Its current distinctive feature is the use of artificial intelligence, machine learning, and reasoning technology to process and analyze big data, effectively reveal the interrelationships between data, and assist scientists in solving the "curse of dimensionality" problem, thereby understanding complex phenomena more quickly and accurately.

The current "Third Axis Era" sees the convergence of "old," "past," and "new" IT technologies, aligning with Karl Popper's three worlds—physical, mental, and artificial. This era emphasizes a win–win, inclusive approach, reflecting the third wave of globalization, characterized by "Positive Sum" interactions. This represents a broader perspective of the World Model compared to Language Large Models (LLMs) and Large Vision Models (LVMs). These current large model technologies

1.3 Parallel Science: AI4S Based on Parallel Intelligence

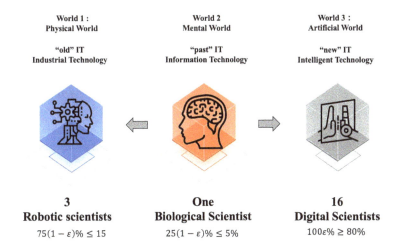

Fig. 1.4 A "day" of parallel scientific research: three worlds, three scientists, three modes

demonstrate that the most direct and natural approach to transforming industries and scientific research is through the parallelization of natural and artificial systems. This includes the transition from natural science to artificial science, from material production to artificial manufacturing, and the integration of digital humans and robots with biohumans in parallel. For the new scientific research paradigm, it is "three worlds, three scientists": "digital scientists," "robotic scientists," and biological (human) scientists, which together constitute parallel scientists (Fig. 1.4).

1.3.2 Digital Scientists and Robotic Scientists

The progress of large models and the rise of prompting engineering suggest that future scientific research will be categorized into "alignment" and "prompting." However, rather than facing unemployment, technology workers in these fields will see a significant increase in their numbers, though their roles may resemble those of "delivery boys." Additionally, there is a shift from "big problems, big models" to "small problems, big models" with vertical segmentation in specific domains. This trend, combined with the continued advancement of large models and the maturation of intelligent agent technology, has naturally led to the emergence of a new type of "digital scientist" focused on "small problems, large models" in scientific research.

At the same time, "robotic scientists" have also been introduced to many scientific research activities beyond digital forms, especially in high-risk, labor-intensive scientific experimental work. For example, the A-Lab [12] has shown the power of robots in accelerating the process of finding new materials. In the near future, from DeSci to self-driving labs and unmanned scientific research factories, "robotic scientists" will become an important component of the "intelligent industry society." The industrialization of scientific research factories is an inevitable trend, and "robotic scientists" will be its key support.

1.3.3 A New "Day" for Parallel Research in Three Modes

Scientific research has progressed from relying on direct observation and experimentation in nature, to conducting experiments in controlled laboratory settings, and now to using mathematical reasoning for calculations and theoretical experiments. The advent of large models enables the use of extensive artificial systems for virtual parallel experiments, surpassing the impact and benefits of traditional computer simulations. This advancement also facilitates numerous "counterfactual experiments" in the social sciences, promoting the integration of "new liberal arts," "new science," and "new engineering."

As a result, the future scientific research model will open a "day" of parallel scientific research in "three worlds, three modes" as shown in Fig. 1.4:

"**Morning**"—autonomous mode (AM), accounting for more than 80%. The main scientific research work will be completed independently by "digital scientists" and "robotic scientists."

"**Afternoon**"—parallel mode (PM), accounting for less than 15%. At this time, biological scientists must intervene and provide guidance for "digital scientists" and "robotic scientists" through remote control or the cloud to complete scientific research projects.

"**Evening**"—expert or emergency mode (EM), accounting for less than 5%. At this time, biological scientists must be the main body, and corresponding scientific research tasks must be completed on-site.

1.3.4 Cornerstones for AI4S in Industry 5.0

In summary, we have built three cornerstones supporting AI4S in the era of Industry 5.0. As shown in Fig. 1.5, the first cornerstone includes big models, scenarios engineering (SE) [13], and human-oriented operating systems (HOOS). The second cornerstone is three types of workers in Industry 5.0: biological workers (approximately 5%), digital workers (approximately 80%), and robotic workers (approximately 15%). The third cornerstone is three kinds of operation modes: autonomous modes (AM, more than 80%), parallel modes (PM, less than 15%), and expert/emergency modes (EM, less than 5%).

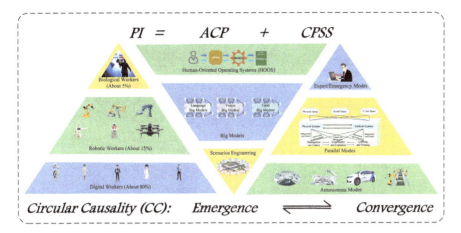

Fig. 1.5 Three cornerstones [11] for AI4S in Industry 5.0

At the same time, AI4S activities can be regarded as a kind of CPSS because they inherently involve the integration of advanced AI algorithms (cyber systems), physical laboratory equipment and sensors (physical systems), and the participation of human researchers and organizational frameworks (social systems). This integration mirrors the core principles of CPSS, which combines cyber, physical, and social components to create intelligent, adaptive systems. Consequently, AI4S not only applies sophisticated AI techniques to scientific research but also aligns with the broader, interdisciplinary approach of CPSS, enhancing the efficiency and effectiveness of research while addressing human and societal needs.

To conclude, the three cornerstones and the ACP approach in Parallel Intelligence will also be essential supports for the development of AI4S. To this end, in the following chapters, we will provide a unique overview of frontiers in AI4S from the perspective of Parallel Intelligence.

1.4 HANOI-AI4S: A General Framework for Parallel Science

1.4.1 Challenges in AI4S

Although AI4S has achieved great advancement in recent years, it still faces challenges, spanning technical, ethical, and practical aspects. As shown in Fig. 1.6, this part gives a brief discussion of the key challenges including the following factors.

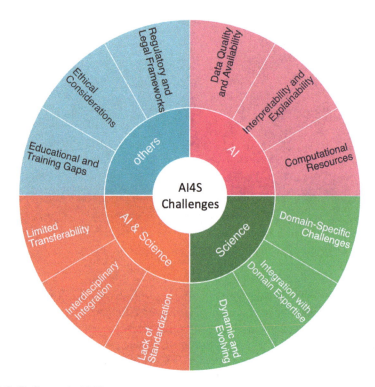

Fig. 1.6 Challenges in AI4S

1.4.1.1 Challenges from AI

Data Quality and Availability Many scientific domains require high-quality and domain-specific data, which may be scarce, incomplete, or noisy. Obtaining and curating reliable datasets for training AI models can be a significant challenge.

Interpretability and Explainability Many AI models, especially deep learning models, are often considered as "black boxes," making it challenging to interpret their decisions. Ensuring interpretability and explainability is crucial, especially in scientific research where understanding the reasoning behind AI-driven insights is essential.

Computational Resources Developing and training complex AI models, especially deep learning models and large language models, requires significant computational resources. Access to high-performance computing infrastructure may be limited, hindering the scalability and deployment of AI4S applications.

1.4 HANOI-AI4S: A General Framework for Parallel Science

1.4.1.2 Challenges from Sciences

Domain-Specific Challenges Different scientific disciplines have unique challenges and requirements. Adapting AI techniques to the specific characteristics of each domain, such as physics, biology, or chemistry, poses challenges in terms of model development and application.

Integration with Domain Expertise Effective collaboration between AI researchers and domain experts is essential. Bridging the gap between technical expertise and domain-specific knowledge is necessary to develop AI4S solutions that align with the needs and constraints of the scientific community.

Dynamic and Evolving Nature of Science Scientific knowledge evolves, and new discoveries can challenge existing models and paradigms. AI4S applications must be adaptable to the dynamic nature of scientific research and able to incorporate new insights.

1.4.1.3 Challenges from Both AI and Science

Lack of Standardization The absence of standardized practices and evaluation metrics in AI4S can lead to difficulties in comparing and reproducing research results. Standardization is crucial for benchmarking, evaluating models, and promoting consistent methodologies.

Interdisciplinary Integration Integrating AI techniques across diverse scientific disciplines poses challenges due to differences in data formats, methodologies, and research paradigms. Bridging these gaps to create a unified AI4S framework requires overcoming disciplinary boundaries.

Limited Transferability AI models developed for one scientific domain may not be easily transferable to another due to differences in data distributions, feature spaces, and underlying processes. Ensuring the generalizability of models across domains is a persistent challenge.

1.4.1.4 Challenges from Other Aspects

Educational and Training Gaps The rapid evolution of AI technologies may lead to gaps in education and training. Ensuring that researchers and practitioners have the necessary skills to understand, implement, and evaluate AI4S methods is crucial for the field's advancement.

Ethical Considerations Ethical challenges in AI4S include issues related to bias in training data, fairness, privacy concerns, and the responsible use of AI technologies. Ensuring that AI applications adhere to ethical standards is crucial for maintaining trust in the scientific community.

Regulatory and Legal Frameworks Developing and deploying AI4S applications requires navigating complex regulatory and legal landscapes. Ensuring compliance with regulations, addressing data privacy concerns, and managing intellectual property rights are ongoing challenges.

Addressing these challenges requires collaborative efforts from researchers, policymakers, and practitioners in both the AI and scientific communities. As the field continues to mature, overcoming these challenges will be essential for realizing the full potential of AI in advancing scientific research.

1.4.2 The Necessity of a General AI4S Framework

To address the aforementioned challenges, a general and unified framework is necessary and essential for AI4S across multiple disciplines. Benefits of introducing a comprehensive and unified AI4S framework are summarized in Fig. 1.7 and the followings.

Fig. 1.7 Benefits of introducing a comprehensive and unified AI4S framework

1.4 HANOI-AI4S: A General Framework for Parallel Science

Interdisciplinary Collaboration Different scientific disciplines often have unique challenges, data types, and methodologies. A unified framework facilitates interdisciplinary collaboration by providing a common ground for researchers from diverse backgrounds to share insights, methodologies, and best practices.

Knowledge Integration A unified framework enables the integration of knowledge from various scientific domains. AI4S applications can leverage insights and techniques from one discipline to inform and enhance research in another, fostering a more holistic and synergistic approach to problem-solving.

Efficient Resource Utilization Developing a common framework allows for the efficient utilization of resources, in terms of both computational power and expertise. Researchers can leverage shared tools and methodologies, avoiding the duplication of efforts and resources across different scientific domains.

Systematic Knowledge Transfer The development of a common framework facilitates systematic knowledge transfer between AI researchers and domain experts in various scientific fields. This exchange of knowledge is crucial for translating AI advancements into meaningful contributions to scientific research.

Transferability of Models and Methods A unified framework promotes the transferability of AI models and methods across disciplines. Models developed for one scientific domain may find applications in another with minimal modifications, accelerating the development and deployment of AI4S solutions.

Benchmarking and Evaluation A common framework provides a basis for benchmarking and evaluating AI4S methods. Standardized metrics and evaluation criteria allow researchers to assess the performance of models consistently, facilitating comparisons and advancements in the field.

Addressing Common Challenges Many scientific domains face common challenges, such as data scarcity, noise, and interpretability issues. A unified framework allows researchers to address these challenges collectively, developing generalizable solutions that can benefit multiple disciplines.

Cross-Domain Insights A shared framework encourages the exploration of cross-domain insights. Researchers can identify common patterns, relationships, and principles that may not be apparent when focusing solely on a single scientific discipline, leading to new discoveries and innovations.

Facilitating Education and Training A unified framework simplifies education and training for researchers entering the AI4S field. Standardized tools and methodologies make it easier for individuals with expertise in one discipline to apply AI techniques in another, promoting a more inclusive and collaborative research environment.

Guiding Ethical Considerations A unified framework provides a foundation for addressing ethical considerations and standards in AI4S. Shared guidelines can help researchers navigate ethical challenges, ensuring responsible and transparent use of AI technologies in scientific research.

Promoting Standardization Standardization in AI4S, facilitated by a unified framework, promotes consistency and reproducibility in research. Standardized practices contribute to the credibility of AI applications and encourage the adoption of best practices across disciplines.

In summary, a general and unified framework in AI4S promotes collaboration, knowledge integration, and efficient resource utilization across multiple scientific disciplines. It provides a foundation for addressing common challenges, fostering cross-disciplinary insights, and guiding the ethical and responsible use of AI technologies in scientific research.

1.4.3 HANOI-AI4S: A General Framework for AI4S

The new philosophy of intelligence will transform our worlds into "6S" societies with "6I": safe in the physical world, secure in the cyber world, sustainable in the ecological world, sensitive to individual needs, serves for all, and smart in all, with cognitive intelligence and parallel intelligence for intelligent science and technology, crypto intelligence and federated intelligence for intelligent operations and management, and social intelligence and ecological intelligence for smart development and sustainability. To this end, Prof. Fei-Yue Wang proposed the HANOI approach [3, 14], integrating Human, Artificial, Natural, and Organizational Intelligence, to realize knowledge automation for sustainable and smart societies.

Aiming for a general framework for AI4S, we propose Hanoi-AI4S, leveraging the HANOI approach with enabling technologies such as digital twins, metaverses, Web 3.0, and blockchain.

The Hanoi-AI4S framework is featured as a multidimensional approach that not only aids in the systematic analysis of AI4S but also facilitates cross-disciplinary comparisons. In addition, it fosters a collaborative environment where insights and innovations can be shared across traditionally isolated scientific communities.

Hanoi-AI4S will serve as a foundational structure for dissecting AI4S components across dimensions such as Real Scientific Problems (Natural world), Artificial Systems, Domain Knowledge, Dataset, Human Roles, and Organizational Mechanism (e.g., DAO and DeSci), and AI Methods in the pipeline of prediction and prescription, as shown in Fig. 1.8.

Artificial System Artificial systems play a crucial role in the Parallel Intelligence framework, providing a test-bed for real scientific problems and enabling highly efficient computational experiments. In AI4S, many works utilize simulators for various purposes, such as data/instruction generation and result validation.

An artificial system is not limited to virtual realities or digital twins. Recently, with the rise of generative AI methods, foundational models like Large Language Models (LLMs) and multimodal models have actually functioned as virtual systems

1.4 HANOI-AI4S: A General Framework for Parallel Science

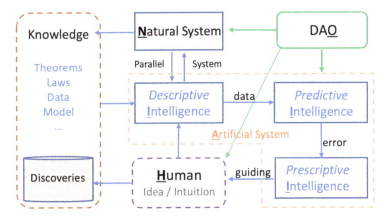

Fig. 1.8 A general framework for AI4S based on Parallel Intelligence, integrating Human, Artificial, Natural, and Organizational Intelligence to realize knowledge automation

of the real world (or world models), demonstrating significant potential in research areas such as proving mathematical conjectures and discovering new materials.

Domain Knowledge Domain Knowledge is essential for AI4S because it provides the contextual understanding and expertise required to effectively apply AI methods to specific scientific problems. For example, in physics and biology, where the intricacies of natural phenomena are vast and varied, having domain knowledge is crucial for designing accurate models, interpreting results, and making meaningful contributions to scientific understanding.

Dataset Data gathered from observations, experiments, simulations, and synthesis is essential for AI4S. Data serves as the foundation for training, validating, and improving AI models, enabling them to make informed predictions, discover patterns, and contribute to scientific understanding.

The symbiotic relationship between AI and data is fundamental to unlocking new insights, making predictions, and advancing our understanding of the natural world across various scientific disciplines. In essence, data is the lifeblood of AI4S, providing the empirical basis for understanding scientific phenomena and training models to contribute to scientific knowledge. The richness, quality, and diversity of data directly impact the effectiveness of AI applications in scientific research.

Human Roles Scientists and AI researchers play pivotal roles in the research and development of AI4S. The interdisciplinary partnership between AI specialists and domain experts is essential for the successful integration of AI technologies into scientific research across diverse fields.

While AI algorithms and technologies contribute significantly to scientific endeavors, human scientists bring essential skills, expertise, and critical thinking to guide, interpret, and contextualize AI applications in scientific research.

Meanwhile, the collaboration between human scientists and AI technologies is symbiotic. While AI brings computational power and efficiency, human scientists contribute creativity, expertise, and contextual understanding, ensuring that AI4S applications align with scientific goals and ethical standards.

Organization and Ecosystem As scientific research is a complex project involving much demands on human, equipment, finance supports, etc., the organization and ecosystem are important for AI4S.

An organized ecosystem promotes the sustainability of AI4S efforts by ensuring that resources are used efficiently and effectively. It allows for long-term planning and investment in AI4S projects, ensuring their continued success and impact.

Compared with AI models/algorithms, the organization and ecosystem of AI4S are much less developed. Nevertheless, a good trend is that more and more scientists from various disciplines are proposing to take advantages of new paradigms like DeSci [15, 16], DAO [14, 17], and Federated Intelligence [18]. As a cornerstone of the proposed Hanoi-AI4S framework, we will give a discussion on this topic in the last chapter with more details.

1.4.4 AI Methods Used in AI4S

AI4S encompasses a variety of AI methods that are tailored to address specific challenges and opportunities within scientific research. The selection of AI methods depends on the nature of the scientific problem, the available data, and the goals of the research. Given the breadth and depth of AI techniques, it is challenging to cover all AI methods comprehensively in this booklet. We recommend that readers refer to a textbook or Wikipedia for specific AI methods whenever necessary. As shown in Fig. 1.9, this part only lists several AI methods commonly employed in AI4S.

Machine Learning Supervised Learning has been used for tasks where the algorithm is trained on labeled datasets, such as classification and regression. In

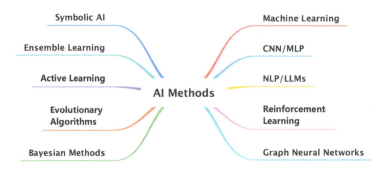

Fig. 1.9 AI methods commonly employed in AI4S

AI4S, supervised learning can be applied to predict outcomes based on historical data, such as identifying disease markers in genomics. On the other hand, Unsupervised Learning is employed for tasks without labeled data, such as clustering and dimensional reduction. Unsupervised learning can reveal hidden patterns in scientific datasets, aiding in the discovery of novel relationships.

In the era of Deep Learning, deep learning models, particularly neural networks, are employed for tasks that involve complex patterns and large datasets. In AI4S, neural networks have been widely used with a variety of structures. Multilayer Perceptron (MLP) and Convolutional Neural Network (CNN) are among most popular models suitable for analyzing spatial data, such as images. For example, in scientific fields like astronomy, CNNs can identify patterns in celestial images or classify objects.

Reinforcement Learning Reinforcement learning has shown powerful capability for optimization and control. Reinforcement learning is applied in situations where an agent learns to make decisions by interacting with an environment. In AI4S, this can be used for optimizing experimental parameters or controlling complex systems.

Natural Language Processing (NLP) NLP techniques, including RNN and LSTM, have been applied to process sequential data. With the introduction of transformer, NLP has become boosting in recent years.

Large Language Models (LLMs) Large Language Models, such as GPT series, have shown great potential in AI4S tasks arranging from summarizing complex texts to identifying key information, inspiring new ideas, etc. They play crucial roles in advancing scientific research by providing insights, automating tasks, and facilitating knowledge discovery.

Evolutionary Algorithms Evolutionary algorithms, including genetic algorithm (GA), are used for optimizing parameters in complex scientific models. These optimization methods are essential for simulations, experimental design, and parameter tuning in machine learning models.

Bayesian Methods Bayesian methods are employed for modeling uncertainty in scientific predictions. They can be used to quantify uncertainty in experimental results or simulations, providing a more comprehensive understanding of the data.

Active Learning Active learning involves an iterative process where the algorithm actively selects the most informative data to label from an unlabeled dataset. Active learning methods in AI4S aim to optimize the learning process, improve model performance, and accelerate scientific discoveries by focusing on the most relevant data points for labeling and analysis.

Ensemble Learning Ensemble methods combine multiple models to enhance overall performance. In AI4S, ensemble learning can be applied to improve the robustness and accuracy of predictions, especially in situations with heterogeneous data sources.

Graph Neural Networks (GNNs) GNNs are used for analyzing complex relationships in networked data, such as biological pathways or social interactions. In AI4S, GNNs can model interactions in molecular structures or study ecological networks.

Symbolic AI Symbolic AI involves representing knowledge using symbols and rules. In AI4S, symbolic reasoning can be applied to logical inference, aiding in hypothesis generation and validation.

The selection of AI methods in AI4S is contingent upon the specific needs of the scientific problem. Frequently, a blend of these methods is utilized to tackle the complex challenges inherent in scientific research spanning diverse disciplines. Subsequent chapters will delve into a more detailed exploration of AI methods, along with their specific applications in AI4S.

References

1. Wang, F.-Y. (2004). Artificial societies, computational experiments, and parallel systems: A discussion on computational theory of complex social-economic systems. *Complex Systems and Complexity Science, 1*(4), 25–35.
2. Li, L., Lin, Y., Zheng, N., & Wang, F.-Y. (2017). Parallel learning: A perspective and a framework. *IEEE/CAA Journal of Automatica Sinica, 4*(3), 389–395.
3. Wang, F.-Y. (2022). Parallel intelligence in metaverses: Welcome to Hanoi! *IEEE Intelligent Systems, 37*(1), 16–20.
4. Wang, F.-Y. (2010). The emergence of intelligent enterprises: From CPS to CPSS. *IEEE Intelligent Systems, 25*(4), 85–88.
5. Wang, F.-Y. (2006). On the modeling, analysis, control and management of complex systems. *Complex Systems and Complexity Science, 3*(2), 26–34.
6. Wang, F.-Y. (2023). New control paradigm for industry 5.0: From big models to foundation control and management. *IEEE/CAA Journal of Automatica Sinica, 10*(8), 1643–1646.
7. Wang, F.-Y. (2004). Computational theory and methods for complex systems. *China Basic Science, 6*(41), 3–10.
8. Wang, F.-Y. (1994). *Shadow systems: A new concept for nested and embedded co-simulation for intelligent systems*. University of Arizona.
9. Wang, F.-Y. (2015). CC5.0: Intelligent command and control systems in the parallel age. *Journal of Command and Control, 1*(1), 107–120.
10. Wang, F.-Y. (2014). Industry 4.0: The queen's new clothes. *Science Times*.
11. Wang, X., Yang, J., Wang, Y., et al. (2023). Steps toward industry 5.0: Building "6S" parallel industries with cyber-physical-social intelligence. *IEEE/CAA Journal of Automatica Sinica, 10*(8), 1692–1703.
12. Szymanski, N. J., Rendy, B., Fei, Y., Kumar, R. E., et al. (2023). An autonomous laboratory for the accelerated synthesis of novel materials. *Nature, 624*, 86–91.
13. Li, X., Wang, K., Tian, Y., Yan, L., Deng, F., & Wang, F.-Y. (2022). From features engineering to scenarios engineering for trustworthy AI: I&I, C&C, and V&V. *IEEE Intelligent Systems, 37*(4), 18–26.
14. Miao, Q., Zheng, W., Lv, Y., Huang, M., Ding, W., & Wang, F.-Y. (2023). DAO to HANOI via DeSci: AI paradigm shifts from AlphaGo to ChatGPT. *IEEE/CAA Journal of Automatica Sinica, 10*(4), 877–897.
15. Hamburg, S. (2021). Call to join the decentralized science movement. *Nature, 600*(7888), 221–221.

16. Wang, F.-Y., Ding, W., Wang, X., Garibaldi, J., Teng, S., Imre, R., & Olaverri-Monreal, C. (2022). The DAO to DeSci: AI for free, fair, and responsibility sensitive sciences. *IEEE Intelligent Systems, 37*(2), 16–22.
17. Wang, F.-Y., Lin, Y., Ioannou, P. A., Vlacic, L., et al. (2023). Transportation 5.0: The DAO to safe, secure, and sustainable intelligent transportation systems. *IEEE Transactions on Intelligent Transportation Systems, 24*(10), 10262–10278.
18. Wang, F.-Y., Qin, R., Chen, Y., et al. (2021). Federated ecology: Steps toward confederated intelligence. *IEEE Transactions on Computational Social Systems, 8*(2), 271–278.

Chapter 2
AI for Mathematics

2.1 Guiding Mathematical Intuition with AI

One of the key drivers of mathematical progress is the exploration of models and the formulation of useful conjectures, which are statements believed to be true but not yet proven universally. Throughout history, mathematicians have utilized data in this process, from early manual computations like Gauss's work on prime number tables that influenced the discovery of the Prime Number Theorem, to modern computer-generated data used in proving conjectures such as the Birch and Swinnerton-Dyer conjecture.

Since the 1960s, computers have been instrumental in generating and analyzing data, leading to insights into previously unsolved problems. Despite significant advancements aided by computer technology, recent progress appears to have plateaued. However, artificial intelligence, particularly the rapid evolution of machine learning and deep learning, has introduced new techniques for identifying patterns in data. In mathematics, AI has proven invaluable for identifying counterexamples to conjectures, identifying structures in mathematical objects, and more.

The article "Advancing mathematics by guiding human intuition with AI" [1] introduces examples of mathematicians discovering new conjectures and theorems in foundational areas of mathematics with the assistance of machine learning. The intuition of mathematicians plays an extremely important role in mathematical discoveries, as "It is only with a combination of both rigorous formalism and good intuition that one can tackle complex mathematical problems." The AI for mathematics methods described in this article is based on the core idea of using machine learning to guide the intuition of mathematicians and propose conjectures, as shown in Fig. 2.1.

To achieve this, the paper describes a general framework through which mathematicians can use tools from machine learning to guide their intuition about complex mathematical objects, verify their assumptions about possible relationships, and help understand these relationships. With the help of this framework, it demonstrates

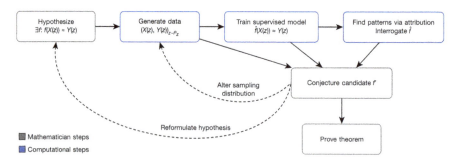

Fig. 2.1 The framework to guide a mathematician's intuition about a hypothesized function f. The process is iterative and interactive in a closed form. Original figure refers to Fig.1 in "Advancing mathematics by guiding human intuition with AI"[1]

successful application in two fields of foundational mathematics. In each case, it shows how the framework makes meaningful mathematical contributions to important open problems: new connections between algebraic and geometric structures, and candidate algorithms predicted by the conjecture of combinatorial invariance of symmetric groups.

2.1.1 Knowledge: Topology

This chapter places high demands on expertise in specific areas of mathematics. Specifically, researchers must master the basic knowledge of topology and the basic properties of knots. On this basis, they can combine mathematical conjecture problems with machine learning methods.

Knots, as simple closed curves in R^3, are central objects of study in low-dimensional topology. The primary goals of this field include classifying knots, understanding their properties, and establishing connections with other areas of mathematics. Invariants play a key role in achieving these goals, as they are algebraic, geometric, or numerical quantities that remain the same for equivalent knots. This chapter focuses on two categories of invariants: hyperbolic invariants and algebraic invariants. These two types of invariants stem from distinct mathematical disciplines, making their connection a topic of significant interest.

2.1.2 AI: Neural Networks

A relatively basic AI model was used in this chapter, that is, a fully connected feed-forward neural network with multiple hidden layers, using the sigmoid function as the activation function. The task is constructed as a multi-class classification

2.2 FunSearch: Searching New Programs for Combinatorial Optimization

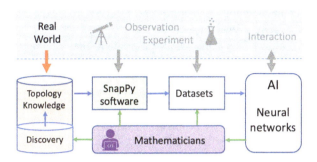

Fig. 2.2 A HANOI-AI4S perspective for discovering new conjectures in topology. It is a mathematician-oriented framework to guiding the intuition for new conjectures

problem, with different signature values as categories, cross-entropy loss as the optimizable loss function, and test classification accuracy as the performance metric.

The framework aids mathematicians in two key ways: first, by confirming the hypothesized presence of structure or patterns in mathematical objects using supervised machine learning, and second, by facilitating comprehension of these patterns through attribution techniques.

A number of datasets were generated from different distributions on the set of knots using the SnapPy software package. The dataset includes: (1) All knots up to 16 crossings (1.7×10^6 knots), taken from the Regina census, (2) Random knot diagrams of 80 crossings generated by SnapPy's *random_link* function (10^6 knots), and (3) Knots obtained as the closures of certain braids.

2.1.3 Human Role: Mathematician in the Loop

The framework proposed in this chapter is a system centered on scientists (specifically, mathematicians), as shown in Fig. 2.2. The role of the mathematician in this process is to guide the selection of conjectures that not only fit the data but also look interesting, appear to be correct, and, ideally, suggest a proof strategy.

2.2 FunSearch: Searching New Programs for Combinatorial Optimization

In general, pure mathematical problems like Extremal Combinatorics may not have direct applications in the real world. However, they often lead to the development of mathematical techniques, algorithms, and insights that can be applied in practical ways. Researchers and mathematicians often explore such problems to deepen their understanding of mathematical structures and to uncover unexpected connections between different areas of mathematics.

The "cap set problem" is a mathematical problem that falls under the domain of Extremal Combinatorics. Specifically, it is related to finite geometry and the study of subsets of vector spaces over finite fields. The problem involves finding sets of vectors with a special property in these finite spaces. The terminology "cap set" refers to a set of vectors that avoids certain additive patterns. Among pure mathematical problems, "cap set problem" is attracting wide attentions including the well-known mathematician Terence Tao.

The "cap set problem" has remained open for a significant amount of time, and finding optimal solutions or establishing bounds has proven elusive. The difficulty is mainly attributed to its connections with deep mathematical concepts and the complexity of the underlying structures involved. The cap set problem is part of a class of problems known to be computationally hard. There were no known polynomial-time algorithms for solving the cap set problem in its full generality.

In "Mathematical discoveries from program search with large language models" [2], the authors proposed FunSearch, through which new constructions of large "cap sets" are discovered, going beyond the best-known ones.

FunSearch is based on a pretrained Large language model (LLM) paired with a systematic evaluator, running in an iterative evolutionary procedure.

In addition to cap set problem, the FunSearch also shows its generality in solving the "online bin packing problem." The "bin packing problem" is a classic Combinatorial Optimization problem in the field of computer science and mathematics. The basic idea is to efficiently pack a set of items of different sizes into a minimum number of containers or bins, each with a fixed capacity. The goal is to minimize the number of bins used while ensuring that the total size of the items in each bin does not exceed its capacity. This problem is known to be NP-hard, meaning that there is no known polynomial-time algorithm to solve it optimally in all cases.

In contrast to the "cap set problem," the "bin packing problem" has numerous real-world applications in logistics, resource allocation, and scheduling. Some examples include: shipping and container loading, scheduling and time management, resource allocation, etc. Solving the "bin packing problem" in real-world scenarios can lead to cost savings, improved efficiency, and better resource utilization, making it a relevant and practical optimization challenge in various industries.

However, the bin packing problem is challenging due to its computational complexity, combinatorial nature, and its applications in real-world scenarios where finding an optimal solution may not be practical within reasonable time frames. Researchers often explore heuristic and approximation algorithms to address these challenges and find effective solutions in practice.

Although some heuristics (and their variants) for solving bin packing problem have strong performance in worst case, they perform poorly in practice. Instead, the first fit and best fit are the most commonly used heuristics. With the help of LLM and evolutionary method, FunSearch discovers better heuristics than first fit and best fit.

2.2.1 *Extremal Combinatorics and Combinatorial Optimization*

Extremal combinatorics is a branch of mathematics that focuses on investigating the maximum or minimum sizes of combinatorial structures satisfying specific conditions. Key domains within extremal combinatorics include extremal graph theory, Ramsey theory, hypergraph Turán problems, set systems and families, codes and designs, combinatorial geometry, random graphs, combinatorial optimization, probabilistic methods, additive combinatorics, and Erdős problems. Researchers in this field analyze the extremal properties of graphs, hypergraphs, set systems, and other combinatorial structures, aiming to optimize or minimize certain parameters while exploring relationships between different mathematical objects. Integrating knowledge from these diverse fields to approach the problem is a challenging task.

Combinatorial optimization involves finding the best solution from a finite set of possibilities for problems characterized by discrete variables and a combinatorial structure. Key domain knowledge in combinatorial optimization includes graph theory, where problems like the traveling salesman problem and graph coloring are studied; linear programming, which deals with optimizing linear objective functions subject to linear constraints; network flow problems, such as the maximum flow and minimum cut problems; integer programming, focusing on optimizing over integer variables; and algorithmic techniques like dynamic programming and greedy algorithms commonly used to develop efficient solutions. Additionally, knowledge of specific problem domains, such as vehicle routing, job scheduling, and facility location, is crucial for applying combinatorial optimization techniques to real-world scenarios and addressing practical challenges in diverse fields like logistics, telecommunications, and manufacturing.

2.2.2 *AI: Genetic Programming and LLM*

As both the "cap set problem" and the "bin packing problem" are NP-hard in complexity, FunSearch does not pursue to solve the problems exactly, instead, it turns to evaluate how good an approximate solution much more efficiently. Specifically, FunSearch tries to generate "program" to solve the problems with high evaluated score. The FunSearch pipeline comprises three key factors, as shown in Fig. 2.3:

(1) Genetic Programming (GP) to generate optimal "program." GP is an evolutionary algorithm-based methodology inspired by biological evolution to evolve computer programs that perform a specific task or solve a particular problem. It falls under the broader category of genetic algorithms, which are optimization algorithms based on the principles of natural selection and genetics. Genetic programming is a versatile approach and has been applied to various problem

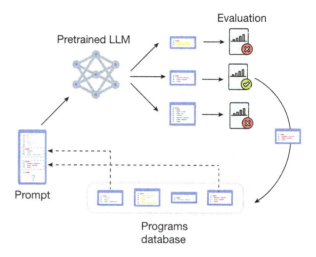

Fig. 2.3 Principle of FunSearch. Full figure refers to Fig.1 in "Mathematical discoveries from program search with large language models" [2]

domains, including symbolic regression, symbolic expression discovery, automatic code generation, etc. The evolution in genetic programming mimics the principles of natural selection, crossover, and mutation, allowing the algorithm to explore a large search space and discover solutions that can be complex and nonintuitive.

While powerful, genetic programming may require careful tuning of parameters, and the results obtained can depend on the choice of representation, genetic operators, and other algorithmic parameters. It is nontrivial to find a setting of genetic parameters for a specific problem.

(2) Large Language Model (LLM) to set parameters in genetic programming. FunSearch takes advantages of the generative ability of a pretrained LLM.

An LLM refers to a natural language processing model that has been trained on a vast amount of text data before being fine-tuned for specific tasks. These models are typically based on deep learning architectures, and they exhibit the ability to understand and generate human-like language. The most notable examples of pretrained LLMs include OpenAI's GPT (Generative Pretrained Transformer) series, such as GPT-3.

The model is pretrained on a large corpus of diverse text data, which can include books, articles, websites, and other sources. The idea is to expose the model to a wide range of language patterns and contexts. During pretraining, the model learns to predict the next word in a sentence or fill in missing words in a context. This is done in an unsupervised manner, meaning that the model does not have explicit labels for the training data; it learns solely from the patterns present in the text. Most pretrained LLMs, including GPT models, are based on the Transformer architecture. Transformers are particularly well-suited for processing sequential data like language due to their attention mechanisms.

LLMs have exhibited impressive language-related capabilities, including Semantic Understanding, Entity Recognition, Coherent Text Generation,

Question–Answering, Language Translation, Text Summarization, Conversational Agents, etc. But the most important abilities related to FunSearch including:

Text Generation, i.e., LLMs are capable of generating human-like, coherent, and contextually relevant text. They can complete sentences, paragraphs, or even generate longer pieces of content based on input prompts.

Creative Writing, i.e., LLMs can be used for creative writing tasks, including poetry generation, storytelling, and imaginative content creation.

Code Generation (Programming Tasks), i.e., LLMs can assist in code generation tasks. Given a prompt or a description of a programming task, they can generate code snippets or complete programs.

However, it is important to note that while LLMs demonstrate impressive capabilities, they also have limitations, such as potential biases in the training data, sensitivity to input phrasing, and occasional generation of incorrect or nonsensical outputs.

(3) A systematic evaluator to guide the LLM.
Programs generated by the LLM are evaluated and scored on a set of inputs. Programs that were incorrect are discarded, and the remaining scored programs are then sent to the programs database. The programs database keeps a population of correct programs, which are then sampled to create prompts. An islands model is adopted to encourage diversity.

FunSearch does not require any data/corpus other than the publicly available OR-Library bin packing benchmarks.

2.2.3 Human Role: Guiders in Prompt Engineering

Designing effective prompts is crucial for obtaining the desired results from LLMs. A prompt for a large language model is a written or spoken input provided to the model to elicit a desired response. The prompt serves as a way to communicate a task or request to the language model, guiding its behavior and shaping the generated output.

However, designing prompts is nontrivial with a number of considerations, e.g., ensuring the prompt is clear and precisely conveys the task or question you want the model to address, including relevant context in the prompt to guide the model's understanding, testing prompts of different lengths to see how the model responds, using controlled vocabulary and specific terms to convey your instructions, including keywords or key phrases related to the task to guide the model's attention, and so on. In one word, prompt engineering is essential in guiding an LLM to generate good responds in practice.

For FunSearch, emphasis is laid on providing examples or sample inputs to illustrate the desired format or type of response. This can help the model understand the task and produce more accurate results. To be specific, FunSearch works best with an initial "solve" program in the form of a skeleton, in which the FunSearch only has to evolve the critical part that governs its logic.

Prompt engineering is usually labor-intensive. To design the prompts in the form of a skeleton, it is required that the researchers have to design the initial program and identify the critical part left for the LLM to evolve further on. In addition, experimenting with different prompts and iteratively refining them based on the model's responses are also a human-oriented process.

2.3 AlphaGeometry: Proving Theorems for Euclidean Plane Geometry

The history of proving mathematical theorems using AI methods spans several decades and has witnessed significant milestones. The field of automated theorem proving emerged in the 1950s and 1960s, with early efforts focused on developing computer programs for logical reasoning. Symbolic computation systems like MACSYMA and REDUCE in the 1960s contributed to automated reasoning and the simplification of mathematical expressions. Formal methods and proof assistants, such as HOL and Coq in the 1980s and 1990s, provided tools for constructing and verifying formal proofs. Notable achievements include the computer-assisted proof of the Four-Color Theorem in 1976 and the formalization of the Kepler Conjecture in 2017 using the HOL Light proof assistant. In 2017, Google's DeepMind introduced DeepMath, showcasing the potential of neural network-based systems in understanding and proving mathematical theorems. OpenAI's GPT-3, released in 2020, demonstrated the application of large language models in mathematical tasks. While early efforts focused on formal verification in specific domains, recent advancements explore the broader application of AI, machine learning, and language models in mathematical reasoning.

The challenge of theorem proving for learning-based methods arises from the scarcity of training data comprising human proofs translated into machine-verifiable languages, particularly in mathematical domains. Notably, geometry, a domain prominent in Olympiad competitions, presents a unique difficulty due to the limited availability of proof examples in versatile mathematical languages such as the Lean. This scarcity stems from the intricate translation challenges specific to geometry. While geometry-specific languages exist, they are narrowly defined and often fail to capture the complexity of human proofs that employ tools beyond the realm of geometry, such as complex numbers. Consequently, there is a substantial data bottleneck, impeding the advancement of geometry-related theorem proving compared to other areas benefitting from human demonstrations. As a result, current methodologies in geometry heavily rely on traditional symbolic methods and human-designed, hard-coded search heuristics.

To overcome the challenges, experts from Google DeepMind and New York University proposed AlphaGeometry [3], which is a theorem prover for Euclidean

Solution

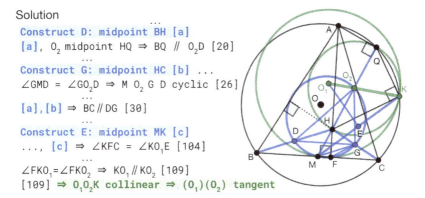

```
Construct D: midpoint BH [a]
[a], O₂ midpoint HQ ⇒ BQ // O₂D [20]
    ...
Construct G: midpoint HC [b] ...
∠GMD = ∠GO₂D ⇒ M O₂ G D cyclic [26]
    ...
[a],[b] ⇒ BC // DG [30]
    ...
Construct E: midpoint MK [c]
..., [c] ⇒ ∠KFC = ∠KO₁E [104]
    ...
∠FKO₁=∠FKO₂ ⇒ KO₁ // KO₂ [109]
[109] ⇒ O₁O₂K collinear ⇒ (O₁)(O₂) tangent
```

Fig. 2.4 Example proof of an IMO problem solved by AlphaGeometry. Full figure refers to Fig.1 in "Solving olympiad geometry without human demonstrations" [3]

plane geometry that sidesteps the need for human demonstrations. AlphaGeometry represents a neuro-symbolic system that employs a neural language model. This model is trained from the ground up using an extensive set of synthetic data. The primary objective of this system is to navigate a symbolic deduction engine through intricate problem-solving scenarios characterized by infinite branching points. Through the fusion of neural and symbolic approaches, AlphaGeometry aims to enhance problem-solving capabilities, particularly in addressing complex challenges with multiple branching decision paths.

AlphaGeometry is highlighted by producing human-readable proofs, substantially outperforming the previous state-of-the-art geometry-theorem-proving computer program, as shown in Fig. 2.4. On a test set of 30 classical geometry problems translated from the IMO, AlphaGeometry approaches the performance of an average International Mathematical Olympiad (IMO) gold medalist.

2.3.1 IMO Euclidean Geometry

To successfully tackle theorem proving problems in Euclidean plane geometry at the International Mathematical Olympiad (IMO), participants require a profound understanding of key geometric principles and a mastery of Euclidean geometry concepts.

First, this encompasses fundamental topics such as angles, triangles, circles, and polygons, along with a deep knowledge of properties, theorems, and geometric constructions. A solid foundation in conic sections, including circles, ellipses, parabolas, and hyperbolas, is indispensable, as is the ability to recognize and apply similarity and congruence theorems. Competitors should also be adept at using trigonometric concepts in a geometric context and understanding geometric transformations, such as reflections and rotations.

Further, the auxiliary construction is essential for theorem proving problems in Euclidean plane geometry because it provides a strategic and often necessary approach to unraveling the complexities of geometric relationships and establishing the validity of conjectures. The introduction of auxiliary constructions involves the creation of additional geometric elements, points, lines, or shapes, to facilitate a clearer understanding of the problem at hand and to establish a path toward a solution. By strategically introducing these constructions, mathematicians can simplify complex configurations, identify hidden symmetries, and create relationships that aid in the logical progression of the proof. Auxiliary constructions serve as a tool to bridge the gap between the given conditions of a problem and the desired conclusion, helping to unveil the underlying geometric structure and guiding the proof toward a successful resolution. Their judicious use is often a hallmark of elegant and insightful solutions in Euclidean geometry, allowing mathematicians to navigate intricate scenarios, leverage known theorems, and ultimately construct a coherent and compelling argument that validates the proposition at stake.

Furthermore, participants should be well-versed in various proof techniques, such as direct proofs and proof by contradiction, to construct clear and rigorous mathematical proofs. Historical theorems and techniques, such as famous results and their proofs, may provide valuable insights. Overall, a combination of geometric intuition, analytical skills, and creative problem-solving approaches is essential for success in tackling the challenging Euclidean geometry problems presented at the IMO.

2.3.2 AI: Symbolic Deduction and Language Model

Symbolic Deduction

Symbolic deduction is a process of reasoning or inference in which conclusions are derived based on the manipulation and analysis of symbolic representations or formal symbols. It involves applying logical rules to symbols to draw logical inferences, deducing new information from existing knowledge. Symbolic deduction is a fundamental concept in logic and artificial intelligence, and it plays a crucial role in areas such as automated theorem proving, formal verification, and knowledge representation.

In the context of mathematical logic, symbolic deduction involves manipulating mathematical expressions, statements, or formulas according to logical rules, such as modus ponens or resolution, to derive new statements or conclusions. This process is often used in the formalization of mathematical proofs and reasoning.

The symbolic deduction process typically involves the following steps: (1) Representation: Expressing knowledge, facts, and rules in a symbolic form using formal symbols or language. (2) Inference Rules: Applying logical rules or inference rules to manipulate symbolic representations and draw conclusions. (3) Derivation: Deducing new symbolic statements or facts based on the application of inference rules. (4) Logical Consistency: Ensuring that the derived conclusions are logically

2.3 AlphaGeometry: Proving Theorems for Euclidean Plane Geometry

consistent and follow from the initial set of symbols and rules. Automated theorem proving, a subfield of symbolic deduction, focuses on developing algorithms and systems that can automatically prove mathematical theorems or logical statements through symbolic manipulation.

Language Model

A language model is a type of artificial intelligence model designed to understand and generate human-like text by learning the inherent patterns and structures of language. These models can be broadly categorized into traditional statistical language models, based on n-gram probabilities, and neural language models, leveraging deep learning techniques like recurrent neural networks (RNNs) or transformer architectures. Neural language models, exemplified by OpenAI's GPT and BERT, have gained prominence due to their ability to capture complex language patterns and dependencies. Language models exhibit contextual understanding, considering the surrounding context to infer meanings and relationships between words. Their generative capabilities make them versatile for tasks like text completion, summarization, and content generation. Many modern language models undergo pretraining on large datasets, learning general language patterns, and can be fine-tuned for specific applications through transfer learning. Applications of language models span a wide range, including natural language understanding, sentiment analysis, machine translation, chatbots, and question–answering systems. Transformer-based models, like GPT, utilize attention mechanisms to effectively capture long-range dependencies in input sequences. The advancement of language models has significantly impacted natural language processing, enabling sophisticated AI applications to interact with and comprehend human language with increasing nuance and context awareness. The first success in applying transformer language models to theorem proving is GPT-f; however, it depends on a substantial amount of human proof examples and standalone problem statements designed and curated by humans.

Auxiliary construction in geometry corresponds to the external generation of terms, exemplifying the expansive nature of theorem proving's infinite branching factor—an acknowledged challenge across diverse mathematical domains, particularly in proving intricate theorems. In the case of AlphaGeometry, a language model was pretrained on a comprehensive set of synthetically generated data. Subsequently, fine-tuning was employed to emphasize the role of auxiliary construction in the proof search process. Notably, AlphaGeometry delegates the deduction proof steps to specialized symbolic engines, streamlining the utilization of the language model to focus specifically on the intricate task of auxiliary construction within the realm of geometry.

2.3.3 Synthetic Dataset

In the realm of deep learning, synthetic data refers to artificially generated datasets created through algorithms rather than collected from real-world observations. The purpose of synthetic data is to augment or replace real data during the training of machine learning models. It can be generated using methods such as procedural generation, simulation, or rule-based algorithms. Synthetic data is often used in combination with data augmentation techniques, where random transformations are applied to existing real data to create variations. This synthetic data approach is particularly valuable in scenarios where obtaining large amounts of real data is challenging, expensive, or time-consuming. It finds applications in domains like computer vision, natural language processing, and medical imaging. Additionally, synthetic data aids in domain adaptation, helping to create representative datasets that resemble the distribution of the target domain. It also addresses privacy concerns associated with real-world datasets by allowing researchers to generate diverse and representative datasets without compromising individual privacy. Despite its advantages, the effectiveness of synthetic data depends on how well it captures the complexities of the real-world data, and models trained on synthetic data may face challenges in generalizing to real-world scenarios if the synthetic data does not fully represent the diversity of the target domain. Overall, synthetic data serves as a powerful tool to enrich training datasets, enhance model robustness, and overcome limitations associated with data scarcity and privacy issues.

Synthetic data also plays a crucial role in theorem proving, with existing methods relying on expert iteration to generate synthetic proofs for predefined problems. The AlphaGeometry stands out by generating both synthetic problems and proofs entirely from scratch. Unlike other approaches, AlphaGeometry does not depend on curated human-defined conjectures, making it distinct from methods like Aygun et al.'s use of hindsight experience replay for generating synthetic proofs. Firoiu et al.'s work, similar to AlphaGeometry, involves synthetic data generation, but it employs a forward proposer and depth-first exploration. In contrast, AlphaGeometry utilizes breadth-first exploration to obtain minimal proofs and premises, employing a traceback algorithm to identify auxiliary constructions and introduce new symbols and hypotheses. The uniqueness of AlphaGeometry's approach, not trained on existing conjectures and employing breadth-first exploration with traceback, makes it a promising method for improving theorem proving in machine learning contexts.

AlphaGeometry purely depends on synthetic data generated from three techniques.

The first technique is using existing symbolic engine named DD. In geometry, a symbolic deduction engine is a deductive database that can efficiently derive new statements from premises using geometric rules. Based on a diverse set of random theorem premises, 100 million synthetic theorems and their proofs are extracted, without using any existing theorem premises from human-designed problem sets.

Second, to widen the scope of the generated synthetic theorems and proofs, component that can deduce new statements through algebraic rules (AR) is introduced in deduction engine. In many Olympiad level proofs, AR is essential to perform angle, ratio, and distance chasing.

Foremost and of utmost significance, the introduction of the dependency difference concept is paramount in generating synthetic proofs. These steps involve auxiliary constructions that exceed the capabilities of symbolic deduction engines, which are not inherently suited for such tasks. This phase yields about ten million synthetic proof steps, each requiring the construction of auxiliary points, thus pushing beyond the limits of pure symbolic deduction. In contrast, any neural solver trained on the synthetic data can proficiently learn auxiliary constructions without human demonstrations or preexisting guidance.

2.3.4 Human Role

The scarcity of translated human proofs into machine-verifiable languages poses a challenge for learning-based theorem proving, particularly in geometry. The auxiliary construction poses a further challenge as it involves exogenous term generation, introducing infinite branching points to the search tree. Existing geometry approaches heavily rely on human-designed heuristics and demonstrations due to the limitations of general-purpose and geometry-specific languages. To address the problem, AlphaGeometry presents an alternative method for theorem proving using synthetic data, thus sidestepping the need for translating human-provided proof examples.

Through synthetic data, inefficient steps such as traditional hand-coded templates have been greatly improved. But this does not mean that the human role can be completely ignored. In AlphaGeometry, pretraining and fine-tuning of Language model depends on human guidance. In addition, evaluation of AlphaGeometry outputs also depends on human experts.

2.4 AlphaTensor: Discovering Faster Matrix Multiplication Algorithm

For thousands of years, algorithms have been assisting mathematicians in performing basic computations; however, the process of discovering new algorithms has always been challenging. DeepMind, dedicated to researching how modern AI technologies can facilitate the automatic discovery of new algorithms, introduced AlphaTensor [4], the first artificial intelligence system capable of discovering novel, efficient, and provably correct algorithms for basic tasks such as matrix multiplication. The core to AlphaTensor is tensor decomposition, and Fig. 2.5 gives a brief illustration.

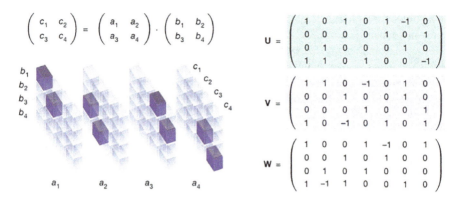

Fig. 2.5 Example of algorithm for matrix multiplication using tensor decomposition. Full figure refers to Fig.1 in "Discovering faster matrix multiplication algorithms with reinforcement learning" [4]

Matrix multiplication is ubiquitous in our daily lives, performing operations behind the scenes in activities such as image processing in smartphones, recognizing voice commands, and generating graphics for computer games. Numerical computation methods used in significant engineering projects, such as finite element methods, also involve matrix multiplication as a core computational task. Companies worldwide are willing to invest substantial time and money in developing computing hardware to effectively address matrix multiplication. Therefore, even small improvements in the efficiency of matrix multiplication can have widespread impacts.

For centuries, mathematicians believed that the standard matrix multiplication algorithm was the most efficient. In 1969, German mathematician Volker Strassen proved that indeed there are better algorithms. For 2×2 matrix multiplication, the Strassen algorithm, compared to the standard algorithm, reduces one multiplication operation from 8 to 7, significantly improving overall efficiency. However, whether more optimal high-order matrix multiplication algorithms exist remains an unresolved challenge. Around 2022, in the field of high-order matrix multiplication algorithms, AlphaTensor has achieved a significant breakthrough in the past 50 years by automating algorithm discovery, unveiling new algorithms that surpass previously known ones.

2.4.1 Knowledge: Matrix Multiplication

Professional knowledge is essential for AlphaTensor in two aspects. The first is that multiplication of matrix A and B can be fully represented by a 3D tensor, as shown in Eq. 2.1.

2.4 AlphaTensor: Discovering Faster Matrix Multiplication Algorithm

$$\mathcal{T}_n = \sum_{r=1}^{R} \mathbf{u}^{(r)} \otimes \mathbf{v}^{(r)} \otimes \mathbf{w}^{(r)}. \tag{2.1}$$

$$\mathcal{S}_t \leftarrow \mathcal{S}_{t-1} - \mathbf{u}^{(t)} \otimes \mathbf{v}^{(t)} \otimes \mathbf{w}^{(t)}. \tag{2.2}$$

The second is the concept of expressing a matrix multiplication bilinear operation, represented by tensor T in the canonical basis, in alternative bases, yielding equivalent tensors; this allows for mapping decompositions obtained in a custom basis back to the canonical basis. To inject diversity into the agent's gameplay, a random change of basis is sampled at the start of each game, applied to T, and AlphaTensor plays the game in that basis, contributing to varied game scenarios.

2.4.2 AI: Deep Reinforcement Learning

In 2016, AlphaGo became the first AI program to defeat the top human players in the game of Go. Subsequently, DeepMind developed AlphaZero, an intelligent agent capable of beating humans in board games such as chess, Go, and shogi. AlphaTensor builds upon the foundation of AlphaZero, marking a significant leap from gaming to scientific research.

Specifically, AlphaTensor transforms the problem of discovering efficient algorithms for matrix multiplication into a single-player game. It utilizes deep reinforcement learning (DRL) to find tensor decompositions within a limited factor space while simultaneously considering the correctness and efficiency of the algorithm (fewer search steps).

It is noteworthy that the search space is exceptionally vast, far exceeding the number of search states in AlphaGo's exploration of Go. To overcome this challenge, AlphaTensor employs a specialized neural network architecture, exploits symmetries inherent in the problem, and leverages synthetic training games.

Data augmentation in tensor factorization training involves extracting additional pairs from played games by leveraging the order invariance of factorizations, achieved through swapping a random action with the last action from each completed game.

AlphaTensor converts the search for the optimal algorithm into a matrix decomposition game, and the entire search process is automatically implemented under the MCTS framework. Human work is reflected in the framework design based on professional domain knowledge. As shown in Fig. 2.6, after the process is started, the human is not in the loop and only needs to analyze and confirm the results after the game is over.

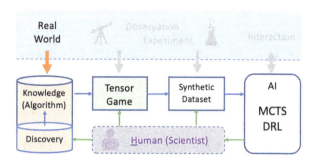

Fig. 2.6 A HANOI-AI4S perspective for discovering faster algorithm of matrix multiplication

2.5 AlphaDev: Discovering Faster Sorting Algorithms

In today's digital society, the daily computing demands of people are constantly increasing, relying on various devices such as computers and smartphones. Over the past fifty years, humans have primarily relied on improvements of processors to meet these demands. However, as chip manufacturing processes approach physical limits, enhancing algorithms running on them to make computations faster and more energy efficient has become crucial. This is particularly important for algorithms that run tens of trillions of times every day, such as sorting algorithms, as they underpin a myriad of data processing applications like online search results and social media post rankings.

In June 2023, DeepMind introduced AlphaDev [5], an artificial intelligence system that utilizes reinforcement learning to discover improved computer science algorithms. Its autonomously developed sorting algorithm surpasses algorithms honed by scientists and engineers over several decades, achieving a maximum performance improvement of 70%. Currently, the algorithms discovered by AlphaDev are included as routines in the libc++ standard sorting algorithm library. These routines are used tens of trillions of times globally each day, making a significant contribution to saving time and reducing energy consumption.

2.5.1 Design and Optimization of Algorithms

Designing the most efficient sorting algorithms, especially when using both assembly language and a high-level programming language like C++, requires a combination of domain-specific knowledge in computer architecture, algorithm design, and programming. Here are key areas of expertise:

Computer Architecture: Understanding the architecture of the target CPU is crucial when working with assembly language. Knowledge of the CPU's instruction set, pipeline structure, cache hierarchy, and memory organization is essential. Awareness of the impact of instruction-level parallelism, pipelining, and cache behavior on the performance of sorting algorithms is also necessary.

2.5 AlphaDev: Discovering Faster Sorting Algorithms

Algorithm Design: In-depth knowledge of sorting algorithms and their time and space complexity is necessary. This includes traditional algorithms like quicksort, mergesort, and heapsort, as well as more recent variations. In addition, familiarity with the strengths and weaknesses of different sorting algorithms under various scenarios, such as data distribution and input size, is also important.

Data Structures: Understanding the characteristics and performance implications of different data structures is essential, as sorting often involves manipulating and comparing elements in various structures (arrays, linked lists, etc.). Knowing how to optimize data access patterns is helpful to enhance cache efficiency.

Assembly Language Programming: Proficiency in assembly language programming for the target architecture is essential. This includes knowledge of registers, addressing modes, and instruction set architecture.

2.5.2 AI: AlphaZero Paradigm

Similar to AlphaTensor, AlphaDev is based on AlphaZero, a learning paradigm that combines deep reinforcement learning and Monte Carlo Tree Search (MCTS) algorithm. AlphaZero has defeated human world champions in various board games such as Go and chess. The success of AlphaDev demonstrates that the AlphaZero paradigm can transition from games to scientific challenges and from simulations to real-world applications.

Code is typically written by people in high-level programming languages like C++, while assembly language is closer to the low-level machine code of computers. To discover more efficient algorithms, AlphaDev transforms sorting into a single-player "assembly game," as shown in Fig. 2.7 with its corresponding perspective based on Parallel Intelligence shown in Fig. 2.8. In this game, players choose a series

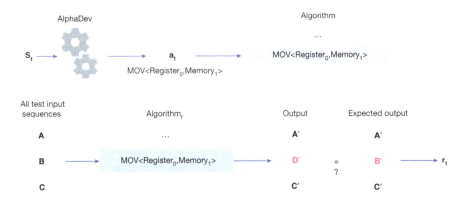

Fig. 2.7 Diagram of AI for faster sorting algorithms. Full figure refers to Fig.2 in "Faster sorting algorithms discovered using deep reinforcement learning" [5]

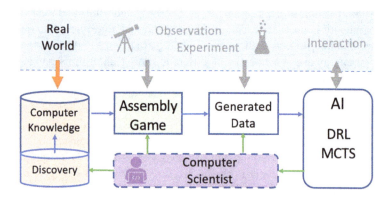

Fig. 2.8 A HANOI-AI4S perspective for faster sorting algorithms

of low-level CPU instructions, referred to as assembly instructions, and combine them to create a new, efficient sorting algorithm.

To play this game, AlphaDev trains an agent to search for correct and efficient algorithms. The agent consists of two core components: (1) a learning algorithm and (2) a representation function. AlphaDev's learning algorithm combines deep reinforcement learning (DRL) and random search optimization algorithms (MCTS), while the representation function captures the underlying structure of assembly programs, primarily using a transformer-based approach. From the perspective of reinforcement learning, the action space consists of a set of instructions, and the reward function is designed to consider both the correctness and computational efficiency (latency) of the algorithm.

This assembly game is highly challenging because AlphaDev must effectively search through a large number of possible instruction combinations to find a sorting algorithm faster than the current best one. The number of possible instruction combinations is similar to the number of possible move combinations in Go. AlphaDev has discovered fixed and variable sorting algorithms from scratch, all of which are novel and more efficient than existing human benchmarks.

2.5.3 Data: Generating Data via Self-playing

AlphaZero employs reinforcement learning. Through trial and error, the program learns to make decisions by receiving feedback in the form of rewards or punishments. AlphaZero combines MCTS with deep neural networks. MCTS is used for decision-making and exploring the game tree, while deep neural networks are employed to evaluate positions and guide the search.

AlphaZero starts without any prior knowledge of the game rules, strategies, or tactics. It learns everything from scratch through reinforcement learning. One key feature of AlphaZero for learning and improving its performance is self-play. It plays games against itself, generating a large-scale dataset to train the neural networks, allowing it to continuously refine its strategies and tactics over time.

A successful algorithm designer in this context would need to bridge the gap between low-level assembly optimization and high-level algorithmic design, considering the intricacies of both domains to achieve the highest efficiency. For innovative approaches like the one described in the AlphaDev example, it is important for a human to have knowledge of reinforcement learning, algorithm design, and optimization techniques. The key role of human is to understand how to model sorting as a reinforcement learning problem and design agents capable of discovering efficient algorithms.

References

1. Davies, A., Veličković, P., Buesing, L., et al. (2021). Advancing mathematics by guiding human intuition with AI. *Nature, 600*(7887), 70–74.
2. Romera-Paredes, B., Barekatain, M., Novikov, A., et al. (2024). Mathematical discoveries from program search with large language models. *Nature, 625*(79395), 468–475.
3. Trinh, T. H., Wu, Y., Le, Q. V., et al. (2022). Solving Olympiad geometry without human demonstrations. *Nature, 625*(79395), 476–482.
4. Fawzi, A., Balog, M., Huang, A., et al. (2022). Discovering faster matrix multiplication algorithms with reinforcement learning. *Nature, 610*(7930), 47–53.
5. Mankowitz, D. J., Michi, A., Zhernov, A., et al. (2023). Faster sorting algorithms discovered using deep reinforcement learning. *Nature, 618*(7964), 257–263.

Chapter 3
AI for Physics

3.1 OmniFold: Unfolding Observables from Large Hadron Collider

Particle physics investigates the fundamental building blocks of matter, addressing questions about their quantity, properties, and interactions. The field employs both theoretical and experimental approaches. Theoretical constraints, such as unitarity and the Pauli exclusion principle, limit the types of subatomic particles that can exist. Experimental verification is crucial due to the existence of multiple self-consistent theories. The Large Hadron Collider (LHC) is a key tool, colliding protons at high speeds to create new particles, as demonstrated by the discovery of the Higgs boson in 2012. However, detecting these particles is challenging, as they are short-lived and often indistinguishable from common background particles. Modern experimental particle physics involves identifying the presence of particles indirectly through their effects.

Particle physics experiments, like those at the LHC, aim to measure properties of particle collisions. Post-detector correction, distributions of collider observables at "truth level" are compared with theoretical predictions, aiding Standard Model understanding and facilitating precision searches for new physics. The critical process of "unfolding" transforms measured information at the detector level into particle-level truth distributions. Unfolding ensures measurement independence from specific experimental contexts, enabling comparisons across experiments and utilizing the latest theoretical tools. However, challenges exist, including the manual determination of measurement binning, limiting the simultaneous unfolding of observables, and insufficient consideration of all auxiliary features controlling detector response. These challenges impact the optimality and potential bias of results, indicating a need for advancements in unfolding methods within particle physics experiments.

OmniFold [1] addresses unfolding challenges in particle physics with a novel approach. As shown in Fig. 3.1, it iteratively unfolds detector-level quantities using

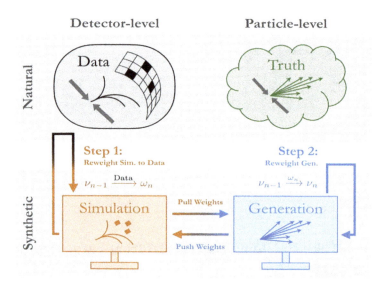

Fig. 3.1 Illustration of OMNIFOLD applied to a set of synthetic and natural data running in parallel. Original figure refers to FIG.1 in paper "OmniFold: A Method to Simultaneously Unfold All Observables" [1]

machine learning, accommodating any-dimensional phase space without binning. This method mitigates issues related to auxiliary features controlling detector response. Unlike previous proposals, OmniFold is suitable for high dimensions and aligns with standard methods in the binned case, naturally handling high-dimensional features akin to past machine learning-based reweighting strategies.

Highlights of OmniFold include: (1) unfolding based on iteratively reweighting a set of simulated events with machine learning, and (2) allowing an entire dataset to be unfolded using all of the available information.

3.1.1 Knowledge: Particle Physics and LHC

Unfolding observables from the Large Hadron Collider (LHC) involves several related domains of knowledge:

Particle Physics: Understanding the properties of particles, their interactions, and the underlying theories (e.g., Standard Model, beyond Standard Model physics) is crucial for interpreting LHC data and unfolding observables.

Quantum Field Theory: Theoretical framework describing particle interactions at the fundamental level, which is essential for modeling and interpreting LHC data.

Monte Carlo Methods: Statistical techniques used to simulate particle collisions and detector responses, which are fundamental for generating synthetic data for training and unfolding.

3.1 OmniFold: Unfolding Observables from Large Hadron Collider 43

Experimental Particle Physics: Knowledge of detector technology, data acquisition, and analysis techniques used at the LHC.

Statistics: Particularly, techniques for data analysis, hypothesis testing, and uncertainty estimation, which are crucial for interpreting unfolded observables.

Computational Physics: Knowledge of numerical methods and algorithms for simulating particle collisions and detector responses, as well as for implementing machine learning models.

High-Energy Physics Phenomenology: Theoretical understanding of particle interactions at high energies, which informs the development and interpretation of LHC experiments.

3.1.2 AI: Parallel Intelligence, Neural Networks, and Neural Resampling

Particle physics presents unique challenges for machine learning compared to other fields. Firstly, quantum mechanics governs particle interactions, leading to quantum interference between signal (e.g., Higgs boson production) and background processes. This means particles can be both signal and background where distributions overlap. Secondly, particle physics benefits from highly accurate simulation tools developed over 40 years. These tools use perturbative quantum field theory to describe particle collisions accurately. However, they produce data in a hundred million-dimensional space, challenging for human or machine visualization. Analysis typically aggregates low-level outputs into composite features, like total particle energy, looking for resonance peaks indicating signals.

Machine learning is well-suited for these challenges, especially when searching for optimal features to discriminate signal from background in vast datasets. The sheer volume of data generated by particle physics experiments, with trillions of recorded and simulated events, requires advanced data processing techniques. Machine learning methods, particularly in high-dimensional spaces, can reveal patterns and extract relevant information from these datasets. The use of synthetic data from simulations further enhances machine learning applications, providing a means to train models on diverse scenarios. In conclusion, the unique characteristics of particle physics, such as quantum interference and complex simulation tools, make it an ideal domain for applying and advancing machine learning methodologies.

The AI methods adopted in OmniFold include three aspects.

First, OmniFold takes neural networks to process jets in their natural representation as sets of particles. Specifically, the architecture of the neural networks is fully connected networks (MLP), named particle flow networks (PFNs).

Second, OmniFold uses neural resampling. In Monte Carlo event generators used in collider physics, the negative weights problem arises due to the nature of the generation process. These generators aim to simulate particle collisions and their outcomes based on theoretical models. However, in some cases, the generated events may have negative weights assigned to them.

Negative weights occur because Monte Carlo simulations use statistical sampling to generate events that are representative of the underlying physics. In certain situations, the sampled events may not accurately reflect the expected distribution of events, leading to a need for adjustments. Reweighing is necessary to address the negative weights problem and ensure that the simulated events accurately represent the physical processes being studied. Reweighing involves adjusting the weights of the generated events so that the overall distribution of events better matches the expected distribution. This adjustment helps to improve the accuracy of the simulation results and ensures that they can be compared effectively with experimental data. In short, the negative weights problem in Monte Carlo event generators arises from discrepancies between the simulated and expected event distributions. Reweighing is necessary to correct these discrepancies and improve the accuracy of the simulation results for data analysis in collider physics.

Traditional reweighting methods involve binning the events and adjusting the weights in each bin to match the expected distribution. However, this approach can lead to loss of information and accuracy, especially in high-dimensional spaces. Neural resampling, on the other hand, uses neural networks to directly learn the reweighting function without binning the events. The neural network takes the characteristics of each event (e.g., particle momenta, energy, etc.) as input and outputs a new weight for the event. This approach allows for a more flexible and accurate reweighting process, as the neural network can capture complex relationships in the data without the need for binning. Neural resampling offers a more efficient and accurate way to perform Monte Carlo reweighting, particularly in high-dimensional spaces where traditional methods may be less effective.

Third, OmniFold takes an iterative reweighting-based unfolding strategy.

Machine learning (ML) has been integral to particle physics for years, employing techniques like recurrent neural networks with b-bagging, the PUMML algorithm, and ResNeXt based on convolutional neural networks. These methods rely heavily on synthetic data for training. While sophisticated, simulations do not always capture subtle correlations exploited by modern ML methods. Historically, there was little need to ensure perfect correlation reproduction. The assumption was that, despite differences, ML should still work. However, this assumption makes assigning uncertainties to ML outputs on real data challenging. Training on real data is an alternative but lacks truth labels. Additionally, in physics, each data point represents both signal and background, making it difficult to use directly for training. Despite these challenges, ML continues to be a powerful tool in particle physics, advancing our understanding of complex datasets.

One approach is training neural networks directly on real data, focusing on particularly clean events for unambiguous labeling. For instance, in particle physics, pairs of top quarks provide clean data points when certain decay patterns are observed. Another method involves training on mixed samples, focusing on learning differences between particles rather than their individual properties. This weakly supervised approach could potentially eliminate the need for simulations in the future. Additionally, fully unsupervised methods like the JUNIPR framework and CaloGAN are being developed for LHC applications. These approaches aim to learn the full data distribution or mimic detector simulations using generative adversarial networks, respectively. Such techniques show promise for advancing machine learning in particle physics, potentially reducing reliance on synthetic data and improving the accuracy of analyses.

Rather than focusing on replicating and generating events similar to particle-level or detector simulations, OmniFold tries to identify areas where simulations are inaccurate. Accordingly, OmniFold takes advantages of neural networks to assess the agreement between simulations and real data, aiming to enhance the simulations. OmniFold introduces a technique where an unsupervised model is trained on synthetic and real data. When disparities arise, the synthetic data is reweighted to match the real data, as shown in Fig. 3.1. OmniFold learns the transformation from simulation to real data, allowing for the inversion of this transformation to effectively eliminate detector simulation effects. This process, known as unfolding, is traditionally tedious and done separately for each observable. As a big leap, OmniFold's ML methods learn how the detector affects each event, enabling the unfolding of any observable using the same network. This approach has the potential to revolutionize experimental analyses, with significant accelerating.

3.2 Magnetic Controlling of Tokamak Plasmas

The tokamak is a type of configuration of nuclear fusion through magnetic confinement, which is a promising way to get sustainable energy. Shaping and maintaining a high-temperature plasma, typically a toroidal (doughnut-shaped) arrangement, using magnetic actuator coils within the tokamak vessel is the key challenge. The precise control of magnetic fields in tokamaks is a complex task, and advanced control systems, feedback mechanisms, and diagnostics are employed to achieve stable and controlled plasma confinement for sustained nuclear fusion reactions. Researchers and engineers continually work on improving magnetic control techniques to advance the development of practical fusion energy.

Research focuses on shaping the plasma distribution in various configurations to optimize stability, confinement, and energy exhaust, especially for experiments. Achieving controlled plasma confinement involves the development of a feedback controller, addressing the tokamak magnetic control problem.

Controllers with feedback mechanism play essential roles in the tokamak magnetic control. The conventional approach involves solving an inverse problem to precompute feed-forward coil currents, followed by the design of single-input single-output PID controllers for stabilizing plasma position and controlling radial position and plasma current. An outer control loop for plasma shape is often added, requiring real-time estimation of plasma equilibrium to modulate feed-forward coil currents. While these controllers are effective, they demand substantial engineering

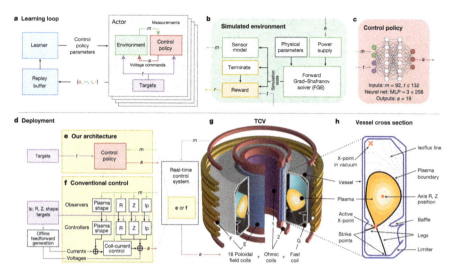

Fig. 3.2 Diagram of AI for magnetic controlling of tokamak plasmas. Full figure refers to Fig.1 in article "Magnetic control of tokamak plasmas through deep reinforcement learning" [2]

effort and expertise, along with complex real-time calculations, especially when changing the target plasma configuration.

To address these problems, scientists from DeepMind and Swiss Plasma Center—EPFL proposed a new nonlinear feedback controller-based reinforcement learning (RL) [2], as shown in Fig. 3.2.

3.2.1 Knowledge: Plasma Physics

The magnetic control of tokamak plasmas requires a combination of expertise in plasma physics, magnetohydrodynamics (MHD), control theory, and specific knowledge related to the behavior of plasmas in a magnetic confinement device. Here are essential domain-specific knowledge areas for effective magnetic control of tokamak plasmas:

Plasma Physics: Understanding the fundamental principles of plasma physics is crucial. This includes knowledge of plasma behavior, instabilities, and the interaction of charged particles with magnetic and electric fields.

Magnetohydrodynamics (MHD): Magnetohydrodynamics is the study of the magnetic properties of electrically conducting fluids, such as plasmas. Expertise in MHD is essential for comprehending the complex interactions between the magnetic field and the plasma in a tokamak.

Magnetic Confinement Theory: Knowledge of magnetic confinement theory is necessary to understand how magnetic fields trap and control the hot plasma. This involves the study of magnetic configurations, stability conditions, and the impact of various perturbations on plasma confinement.

Plasma Equilibrium and Stability: Knowledge of plasma equilibrium and stability is essential for predicting and controlling plasma behavior. This involves understanding the conditions for achieving stable plasma configurations and how perturbations can affect plasma stability.

Knowledge of Tokamak Configurations: Understanding the specific details of the tokamak device being used, including its magnetic geometry, plasma parameters, and any unique features, is essential for effective magnetic control.

Control Theory: Control theory is fundamental for designing feedback control systems to regulate and stabilize the plasma. This includes the application of classical control techniques, such as PID (Proportional–Integral–Derivative) control, as well as more advanced control strategies like model predictive control or adaptive control. Specifically, understanding the dynamics of feedback control systems specific to tokamak plasmas is crucial. This involves knowledge of how measurements from diagnostics are used to adjust magnetic fields in real time to maintain stability and control plasma parameters.

Instrumentation and Diagnostics: Expertise in the instrumentation and diagnostics used to measure plasma properties, magnetic fields, and other relevant parameters is necessary. This includes familiarity with diagnostics like magnetic sensors, Langmuir probes, and other tools used to monitor plasma behavior.

Computational Modeling: Computational modeling skills are important for simulating plasma behavior, predicting the impact of magnetic field adjustments, and optimizing control strategies. This may involve the use of numerical codes and simulations specific to tokamak plasmas.

Real-Time Data Analysis: The ability to analyze real-time data from various diagnostics is crucial for making informed decisions in adjusting magnetic fields to control the plasma. This involves data processing, interpretation, and quick decision-making.

3.2.2 *AI: Reinforcement Learning*

The integration of Reinforcement Learning (RL) into plasma control is a significant advancement, offering an intuitive approach to defining objectives and simplifying control systems. RL's success in various fields focuses on what needs to be achieved, streamlining control systems by replacing nested architectures with a single, efficient controller, and eliminating the need for independent equilibrium reconstruction. These advantages shorten the controller development cycle and accelerate the exploration of alternative plasma configurations. However, when applied to magnetic controller design in fusion control, RL is facing challenges such as high-dimensional measurements, long time horizons, rapid instability growth rates, and inferring plasma shape through indirect measurements.

In [2], the RL algorithm uses simulator data to develop a near-optimal policy based on a specified reward function. Due to the computational demands of evolving the plasma state, the simulator data rate is slower than typical RL environments. To address data scarcity, the algorithm employs Maximum a Posteriori Policy Optimization (MPO), an actor-critic method for efficient data collection across parallel streams. Training involves two neural network architectures: the critic network, incorporating hyperbolic tangent values and an LSTM layer followed by an MLP, and the policy network, including a linear layer, normalization, and a three-layer MLP for Gaussian distribution parameters. Adaptation occurs during training, with only the policy network deployed on the plant, enabling inference within the CPU's L2 cache for compatibility with hardware used for training and achieving a 10-kHz control rate.

Rewards are crucial in Reinforcement Learning, especially in scenarios with multiple objectives. Individual components tracking specific aspects are combined into a unified scalar reward value. Target values for objectives, often time-varying, are integrated into observations for the policy. Shape targets, generated through a shape generator or manual specification, are normalized to 32 equally spaced points along a spline. This consolidation involves computing differences between actual and target values, transforming them into quality measures using nonlinear functions, and merging vector-valued objectives through a weighted nonlinear combiner. The final scalar reward, ranging from 0 to 1, results from a weighted combination of individual quality measures. This stepwise reward is normalized for a maximum cumulative reward of 100 for 1 second of control. In case of policy-triggered termination, a large negative reward is assigned. Quality measures are computed from errors using a softplus or sigmoid, providing a nonzero learning signal in early training stages with large errors. Rewards are combined using a weighted smooth max or geometric mean, encouraging improvement in the worst reward while promoting overall enhancement in all objectives.

Utilizing simulator data, the RL algorithm seeks an optimal policy based on the specified reward function. The dynamics of the plasma, coupled with active and passive conductors, are modeled using the free-boundary simulator, FGE. Within this simulator, synthetic magnetic measurements are generated to emulate TCV sensors, serving as inputs for learning control policies.

3.2.3 Human Role and Collaboration

Magnetic control of tokamak plasmas is a typical research field that requires profound professional domain knowledge. This achievement is the result of the intersection of high-energy physics and artificial intelligence, which can be seen from the collaborators of the paper [2]. DeepMind researchers Jonas Degrave and Jonas Buchli, and Swiss Plasma Center-EPFL researcher Federico Felici contributed to the paper as co-first authors. Scientists at the Swiss Plasma Center—EPFL believe that the collaboration with DeepMind will allow researchers to push the boundaries and accelerate the long journey toward fusion energy.

3.2.4 Summary

This chapter presents a novel method for plasma magnetic confinement in tokamaks, employing a machine learning-based control system. This approach meets community expectations by delivering high performance, robustness in uncertain conditions, intuitive target specification, and unprecedented versatility. Overcoming capability and infrastructure challenges, the study advanced a numerically robust simulator, balanced simulation accuracy with computational complexity, incorporated hardware-specific sensor and actuator models, and trained on realistic variations of operating conditions. Utilizing a highly data-efficient Reinforcement Learning (RL) algorithm scalable to high-dimensional problems, the study employed an asymmetric learning setup with an expressive critic and a fast-to-evaluate policy. Compiling neural networks into real-time-capable code facilitated successful hardware experiments, demonstrating fundamental capability and advanced shape control without plant fine-tuning. Additionally, the study underscores the effectiveness of a free-boundary equilibrium evolution model for developing transferable controllers, suggesting its potential for future device control system testing.

3.3 Finding Evidence for Intrinsic Charm Quarks

Physics textbooks describe protons as subatomic particles containing three quarks bound together by gluons. Protons exist in a state where two up quarks and one down quark are bound by gluons, but quantum theory predicts an infinite number of quark–antiquark pairs in addition to this. The masses of light and heavy quarks are respectively smaller or larger than the proton's mass, and in high-energy collisions, the internal structure of protons is revealed. However, it remains unclear whether heavy quarks, known as intrinsic heavy quarks, are also part of the proton's wave function, a determination made by non-perturbative dynamics and therefore unknown. Specifically, the existence of charm quarks as a substantial component of the proton has been a subject of debate and extensive research, yet a definitive result remains elusive.

The study [3] from The NNPDF Collaboration, utilizing machine learning and a large volume of experimental data, has found evidence for the existence of intrinsic charm quarks through a high-precision determination of the quark–gluon content of nucleons. The momentum distribution of intrinsic charm quarks within three standard deviations is highly consistent with model predictions, and all findings have been validated through comparisons with recent experiments on Z-boson production from the LHCb experiment.

3.3.1 Knowledge: Parton Distribution Functions

Parton Distribution Functions (PDFs) describe the probability density of finding a quark or gluon within a proton, which is crucial for understanding the internal structure of protons and predicting outcomes of high-energy particle collisions. PDFs have played a pivotal role in significant discoveries, including the Higgs boson. However, the limited knowledge of PDFs remains a key constraint in exploring new physics phenomena, e.g., PDFs cannot be computed from first principles, and they must be derived from data through meticulous comparisons between theoretical predictions and experimental outcomes.

3.3.2 AI: Neural Network and Genetic Algorithms

NNPDF is an open-source software framework, and it provides not only a central set of PDFs but also estimates of uncertainties associated with the PDF predictions. We give an illustration of the NNPDF shown in Fig. 3.3. This is important for understanding the reliability of the predictions and their sensitivity to the input data.

NNPDF employs Neural Networks as an unbiased modeling tool, trained with Genetic Algorithms. These networks construct a Monte Carlo representation of PDFs and their uncertainties, forming a probability distribution within a function space. This innovative approach facilitates a comprehensive and data-driven determination of PDFs crucial for advancing our understanding of particle physics.

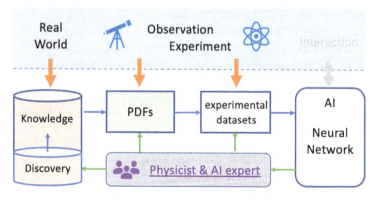

Fig. 3.3 Diagram of AI for finding evidence of intrinsic charm quarks in proton

3.3 Finding Evidence for Intrinsic Charm Quarks

NNPDF uses neural networks to parameterize PDFs, allowing for a more flexible and data-driven approach compared to traditional methods. The basic idea is to use a set of neural networks to model the PDFs, and these networks are trained on a diverse dataset of experimental data from various high-energy experiments. These networks serve as flexible, nonlinear functions that map the relevant input variables to the Parton distribution functions.

The neural networks are trained using the compiled dataset. During training, the networks' parameters are adjusted to minimize the difference between the predicted PDFs and the experimental data.

3.3.3 Collecting Experimental Dataset

NNPDF conducts global fits to a wide range of experimental data from various high-energy physics experiments to determine Parton distribution functions (PDFs) with the highest precision possible. The datasets typically include measurements from experiments that probe the internal structure of protons, combining about 5,000 data points stretching back about 40 years covering about 40 different experiments. Sources of experimental data for NNPDF analyses include but not limited to:

Deep Inelastic Scattering (DIS) Experiments: These experiments involve shooting high-energy electrons or neutrinos at protons and measuring the scattered particles to probe the proton's internal structure.

Drell–Yan Processes: Experiments studying the production of lepton pairs in proton–proton or proton–antiproton collisions, which provide information about the quark and antiquark distributions inside protons.

Electroweak Boson Production: Measurements of processes involving the production of electroweak bosons like W- and Z-bosons, which are sensitive to PDFs.

Heavy Quark Production: Data on the production of heavy quarks (charm and bottom) in various processes, contributing to the understanding of the heavier quark content in protons.

It is important to include a diverse set of experimental measurements covering a broad kinematic range and different processes to obtain a comprehensive understanding of PDFs. The collaboration continually updates its analyses with new experimental data as it becomes available to improve the precision and reliability of the determined Parton distribution functions.

3.3.4 Human Role: The NNPDF Collaboration

It is worth to note that this work is led by a group, i.e., the NNPDF collaboration, instead of several scientists. In the collaboration, scientists play various roles contributing to the overall effort of determining PDFs. The organization of the collaboration involves a coordinated effort with individuals specializing in different aspects.

Theoretical physicists are involved in developing and refining the theoretical framework for the PDF analysis. They work on incorporating quantum chromodynamics (QCD) calculations, perturbative QCD, and other theoretical aspects necessary for predicting Parton distributions.

Scientists with expertise in experimental high-energy physics play a crucial role in providing and interpreting experimental data. Experimentalists contribute to the selection and analysis of data from various experiments, ensuring that the datasets used in the analysis are accurate and relevant.

Professionals with expertise in statistics and data analysis work on developing and applying sophisticated statistical methods to extract PDFs from experimental data. They contribute to the fitting procedure, uncertainty estimation, and optimization of the PDFs.

Given the use of machine learning techniques in NNPDF, Machine Learning Experts in this field are involved in developing and implementing neural network architectures, training algorithms, and optimizing the machine learning aspects of the analysis.

References

1. Andreassen, A., Komiske, P. T., Metodiev, E. M., et al. (2020). OmniFold: A method to simultaneously unfold all observables. *Physical Review Letters, 124*(18), 182001.
2. Degrave, J., Felici, F., Buchli, J., et al. (2022). Magnetic control of tokamak plasmas through deep reinforcement learning. *Nature, 602*(7897), 414–419.
3. The NNPDF Collaboration. (2022). Discovering faster matrix multiplication algorithms with reinforcement learning. *Nature, 608*(7923), 483–487.

Chapter 4
AI for Biology

4.1 AlphaFold: Predicting 3D Protein Structure

Protein folding is a critical process in molecular biology and is essential for understanding the structure and function of proteins. The three-dimensional structure of a protein is closely related to its biological function, and determining these structures experimentally can be time-consuming and expensive. Predicting the three-dimensional structure of a protein from its amino acid sequence, known as the protein folding problem, has been a long-standing challenge in biology and biochemistry. Experimental methods like X-ray crystallography and cryo-electron microscopy have been used to determine protein structures, but these techniques are resource-intensive and may not always be feasible for every protein.

DeepMind introduced AlphaFold [1] and later improved its performance with AlphaFold 2 [2]. As shown in Fig. 4.1, AlphaFold uses deep neural networks to predict the 3D structure of a protein based on its amino acid sequence. The system was trained on a large dataset of known protein structures.

AlphaFold is important for science because it addresses a long-standing problem in biology, providing accurate predictions of protein structures. This has broad implications for understanding biological processes, developing new drugs, and advancing research in fields such as medicine and biochemistry.

Accurate predictions of protein structures have the potential to significantly accelerate research in various scientific and medical fields. Understanding protein structures aids in deciphering their functions, which is crucial for drug discovery, disease understanding, and the development of targeted therapies. AlphaFold's success in the Critical Assessment of Structure Prediction (CASP) competition in 2020 demonstrated its ability to predict protein structures with remarkable accuracy, approaching experimental precision.

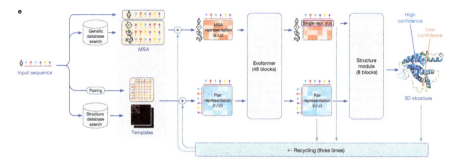

Fig. 4.1 Diagram of AI for predicting the 3D structure of a protein. Full figure refers to Fig.2 in the article of "Highly accurate protein structure prediction with AlphaFold" [2]

4.1.1 Domain Knowledge

Amino Acid Pairwise Potentials: AlphaFold 2 considers amino acid pairwise potentials derived from known protein structures. These potentials provide information about the likelihood of different amino acid pairs being in close spatial proximity in a correctly folded protein.

Multiple Sequence Alignments (MSA): AlphaFold 2 uses information from the evolutionary relationships of proteins, obtained through multiple sequence alignments (MSA). This helps identify conserved patterns and coevolutionary signals between amino acids.

Distance Constraints: AlphaFold 2 uses distance constraints obtained from experimental data, such as cryo-electron microscopy and X-ray crystallography. These constraints provide information about the spatial relationships between pairs of amino acids in a protein.

Integration of Structural Biology Principles: The architecture of AlphaFold 2 is designed to integrate principles from structural biology, capturing the physics and chemistry underlying protein folding. This includes considerations of steric clashes, hydrogen bonding, and other fundamental aspects of protein structure.

4.1.2 AI: Transformer and Attention Mechanisms

AlphaFold and AlphaFold 2 utilize deep learning techniques, specifically deep neural networks, to predict the three-dimensional structure of proteins. The methods involve a combination of convolutional neural networks (CNNs), attention mechanisms, and novel architectures.

AlphaFold uses convolutional neural networks to analyze the MSA data. CNNs are particularly effective in capturing spatial patterns in input data, making them suitable for tasks like image recognition and, in this case, identifying patterns

4.1 AlphaFold: Predicting 3D Protein Structure

in amino acid sequences. AlphaFold 2 introduces attention mechanisms named Evoformer, inspired by the Transformer architecture. Attention mechanisms allow the model to focus on different parts of the input sequence when making predictions. This is crucial for capturing long-range interactions between amino acids in the protein sequence. AlphaFold 2 also introduces an attention-to-distance (A2D) module, which refines the distance predictions made by the model. This module helps the model better capture the geometric relationships between amino acids in the protein sequence.

AlphaFold and AlphaFold 2 are trained end-to-end, meaning the entire system, including all components of the neural network and other modules, is trained jointly. This allows the model to learn complex relationships and dependencies in the data.

4.1.3 Dataset

AlphaFold 2 utilized various datasets for training and validation to develop its predictive model for protein structure.

Protein Data Bank (PDB): The PDB is a primary source of experimentally determined protein structures. AlphaFold 2 likely used a subset of the PDB as a training dataset to learn patterns and features associated with known protein structures.

CASP Datasets: The Critical Assessment of Structure Prediction (CASP) competition provides datasets specifically designed for evaluating and benchmarking protein structure prediction methods. For example, Big Fantastic Database (BFD) is one of the largest publicly available collections of protein families, which consists of 65,983,866 families represented as MSAs and hidden Markov models (HMMs) covering 2,204,359,010 protein sequences from reference databases, metagenomes, and metatranscriptomes [2].

4.1.4 Human Role

As shown in Fig. 4.2, while AlphaFold 2 is an end-to-end deep learning method for protein structure prediction, the role of scientists remains crucial in several aspects of the process.

Data Interpretation: Scientists are responsible for interpreting the predictions generated by AlphaFold 2. They need to assess the reliability and biological relevance of the predicted structures, especially in the context of specific research questions or hypotheses.

Biological Context: Understanding the biological context of protein structures is essential. Scientists provide expertise in the functional implications of protein structures, helping to link the predicted structures to biological functions and pathways.

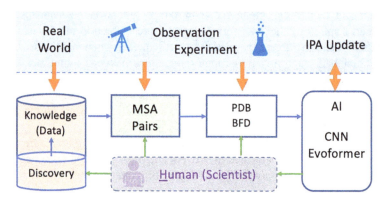

Fig. 4.2 Diagram of AI for predicting the 3D structure of a protein

Experimental Validation: Experimental validation of predicted structures is a critical step. Scientists design and conduct experiments, such as X-ray crystallography or cryo-electron microscopy, to validate and refine the predicted structures provided by AlphaFold 2.

Integration with Experimental Data: Scientists integrate experimental data with AlphaFold 2 predictions to refine and improve the accuracy of the predicted structures. This iterative process ensures that computational predictions align with real-world experimental observations.

Quality Assessment: Scientists assess the quality of predictions and identify potential limitations or uncertainties. They contribute to the development and refinement of metrics for evaluating the accuracy and reliability of predicted protein structures.

Biological Insight: Scientists bring biological insight to the interpretation of predicted structures. They investigate how the predicted structures align with known biology, identify potential functional sites, and explore the implications for disease mechanisms or drug design.

Iterative Modeling: The use of AlphaFold 2 is often part of an iterative modeling process. Scientists may iteratively refine predictions, validate them experimentally, and incorporate new knowledge to continually improve the accuracy of protein structure predictions.

Exploration of Novel Proteins and Complexes: Scientists use AlphaFold 2 to explore the structures of novel proteins and protein complexes. They leverage computational predictions to guide experimental studies on previously uncharacterized biological entities.

Communication of Results: Scientists play a key role in effectively communicating the results of protein structure predictions, including their confidence levels and implications, to the broader scientific community and the public.

As we can see, the involvement of scientists is crucial for contextualizing predictions, validating structures experimentally, and deriving meaningful biological insights. The collaboration between computational methods like AlphaFold 2 and

4.1.5 AlphaFold2 and New Advancements

AlphaFold2 has significantly advanced research in various fields by accurately predicting most protein structures.

Intrinsically disordered regions (IDRs) pose a challenge as they rapidly interconvert between many conformations. AlphaFold2 can identify structures for IDRs that fold under certain conditions, known as conditional folding [3]. This ability helps quantify conditional folding across different species and rationalize disease-causing mutations in IDRs.

Managing the vast number of structural models generated by AlphaFold2 is challenging, especially as the repository of publicly available structures approaches a billion entries. A clustering procedure has been developed to scale to billions of structures, identifying millions of clusters, some of which do not resemble previously known structures [4]. Additionally, combining AlphaFold2 with ARTINA and UCB Shift allows for the automatic assignment of chemical shifts in NMR spectra, enhancing accuracy and robustness for larger systems [5]. Finally, by clustering input sequences, AlphaFold2 can sample multiple biologically relevant conformations of metamorphic proteins with high confidence, advancing our understanding of protein function at the atomic level [6].

The latest version AlphaFold 3 model [7] features a significantly enhanced diffusion-based architecture, enabling it to predict the structures of complexes comprising proteins, nucleic acids, small molecules, ions, and modified residues simultaneously. This new model showcases notably improved accuracy compared to many previous specialized tools, particularly in protein–ligand interactions surpassing state-of-the-art docking tools, protein–nucleic acid interactions surpassing nucleic-acid-specific predictors, and antibody–antigen prediction accuracy.

4.2 RFdiffusion: De Novo Design of Protein

The success of AlphaFold has completely revolutionized the current state of structural biology research. In addition to predicting protein structures, the use of AI for de novo design of functional protein molecules has become a reality. In July 2023, Professor David Baker, a pioneer in the field of artificial protein design, published a latest paper in Nature. His team developed the AI software RFdiffusion [8], which can overcome many previous limitations in protein design. It can "customize" the design of proteins, including high-order symmetrical structures that were previously challenging for AI to design. Proteins designed on-demand by AI models can serve as the foundation for vaccines, therapeutic drugs, and biomaterials, potentially ushering in a transformative era in the field of medical health.

For years, researchers have been striving to construct new proteins. Initially, they attempted to assemble useful portions of existing proteins (such as the bag-like structure in enzymes that catalyze chemical reactions). This method relied on an understanding of protein folding and functioning, along with intuition and extensive trial and error. Scientists sometimes had to sift through thousands of designs to find one that met their expectations. To overcome these challenges, in recent years, scientists from various institutions have developed numerous AI-based protein design tools, including hallucination and inpainting methods. However, these methods fell short of meeting the requirements of protein design. In contrast, RFdiffusion and similar AI protein design methods demonstrate superiority, relying on the latest algorithms of generative artificial intelligence—specifically, denoising diffusion models. Diffusion models are a generative method proven to be useful in image and text generation, such as in popular applications like Stable Diffusion, DALL-E, and Midjourney.

4.2.1 De Novo Design of Protein Structure

The de novo design of protein structure and function requires a combination of expertise in various domains to address the complexity of biological systems and the intricacies of protein engineering. Some key domain-specific knowledge areas include but not limited to the following:

Biochemistry and Molecular Biology: Understanding the principles of protein structure, folding, and function. Knowledge of amino acid properties, interactions, and the relationship between sequence and structure.

Structural Biology: Proficiency in analyzing protein structures using techniques like X-ray crystallography, NMR spectroscopy, or cryo-electron microscopy. Insight into the determinants of protein stability, dynamics, and conformational changes.

Computational Biology and Bioinformatics: Skills in using computational tools and algorithms for protein structure prediction and analysis. Understanding sequence–structure relationships and the prediction of potential binding sites.

Protein Engineering: Knowledge of methods for modifying and designing proteins with specific functions. Familiarity with rational design, directed evolution, and computational protein design approaches.

Immunology and Drug Development: Insight into the principles of immunology for designing proteins with therapeutic applications. Understanding the interactions between proteins and small molecules for drug development.

Biophysics: Knowledge of the physical principles governing protein behavior, including thermodynamics, kinetics, and spectroscopy. Expertise in experimental techniques for characterizing protein properties.

These interdisciplinary skills and knowledge areas are crucial for scientists and researchers involved in the de novo design of protein structures and functions, as

4.2 RFdiffusion: De Novo Design of Protein

they work to create novel proteins with specific and desired properties for various applications, including therapeutics, enzymes, and biomaterials.

4.2.2 AI: Diffusion Model

In the context of artificial intelligence (AI), a diffusion model typically refers to a type of generative model used for simulating the diffusion process of data over time. These models are part of the broader category of generative models that aim to learn and simulate the distribution of data.

The core idea behind diffusion models is to model how a simple distribution transforms into a more complex one over a series of steps or iterations. This process mimics the diffusion of information or features in a dataset. The term "diffusion" is used because the model captures how the data evolves, spreads, or diffuses through the latent space. The diffusion process of RFdiffusion is shown in Fig. 4.3.

RFdiffusion achieves protein structure prediction by fine-tuning the structure prediction network of RosettaFold and integrating it into a denoising diffusion model. After several rounds of noise elimination, it generates a meaningful protein backbone. The protein backbone determines the shape and function of the protein.

Both RosettaFold and RFdiffusion predictions are based on model input conditions, converting coordinates into predicted structures. RosettaFold takes input and output channels of sequence, structural template, and initial coordinates. This enables it to perfectly adapt to the iterative process of "diffusion" with a time scale involving both sequence and structure. In the denoising process, RFdiffusion primarily takes noise coordinates from the previous step as input and, for specific design tasks, can provide a series of auxiliary adjustment information, including partial sequences, folding information, or fixed functional motif coordinates.

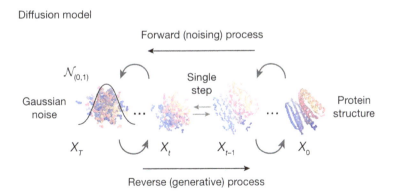

Fig. 4.3 Diagram of AI for de novo design of protein. Full figure refers to Fig.1 in article "De novo design of protein structure and function with RFdiffusion" [8]

4.2.3 Summary

RFdiffusion represents a significant advancement in protein design, outperforming existing methods. It effortlessly produces diverse designs of up to 600 residues, surpassing the complexity and accuracy of previous approaches. Half of the unconditional designs tested exhibit solubility, aligning with design models and displaying high thermostability. Despite increased complexity, RFdiffusion designs match the ideality and stability of prior methods like Rosetta. Electron microscopy confirms structural similarity between the designed oligomers and models, often diverging from known protein structures.

The versatility of RFdiffusion transcends prior achievements, allowing non-experts to generate functional proteins based on minimal specifications, akin to generating images from text prompts. This breakthrough expands the scope of problems addressable with robust and accurate solutions.

While the development of the RFdiffusion model marks a significant step forward in de novo protein design, there are still numerous hurdles to overcome on the path to developing effective therapies. Firstly, there is the challenge of designing more complex binding proteins using the model, such as protein receptors utilized by antibodies or T cells. These proteins possess flexible loop structures that interlock with their targets, whereas the proteins designed by RFdiffusion so far have flat interfaces. Secondly, the RFdiffusion protein design based on the diffusion model cannot create proteins vastly different from natural proteins. This limitation arises from the AI system being trained only on existing protein data, resulting in a tendency to produce proteins with similar structures.

4.3 scGPT: A Foundation Model for Single-Cell Biology Research

Natural language processing (NLP) techniques in AI have been applied to protein research for decades. Protein sequences resemble natural languages, with amino acids forming structures akin to words conveying function. Since 2017, models based on Transformer have advanced text generation to a new level comparable to human-like capabilities. In protein research, dedicated transformers are expected to dominate sequence generation. Fine-tuning these models on protein families will expand their repertoire with potentially functional novel sequences. Controllable design of novel protein functions using tags like cellular compartment or function is feasible. Generative models, including large language models (LLMs) and diffusion models (DMs), show promise in designing functional protein sequences, likely to impact protein design significantly.

Single-cell RNA sequencing (scRNA-seq) has revolutionized our understanding of cellular heterogeneity, disease mechanisms, and potential personalized therapies. It has enabled the creation of comprehensive cell atlases like the Human Cell Atlas, comprising tens of millions of cells. Recent advances in sequencing technology have expanded our capabilities to include epigenetics, transcriptomics, and proteomics, providing multimodal insights. However, this expansion has also led to new challenges such as reference mapping, perturbation prediction, and integrating multi-omics data.

To tackle these challenges, artificial intelligence (AI) is emerging as a promising approach. Current machine learning methods in single-cell research are fragmented, limiting the breadth and scale of datasets used in studies. To address this, a generative pretraining of foundation model named scGPT [9] is proposed. The scGPT model, inspired by self-supervised pretraining in natural language generation and pretrained on over 33 million cells, offers a unified framework for nonsequential omics data. It adapts the transformer architecture to learn cell and gene representations simultaneously, providing fine-tuning pipelines for diverse tasks. This approach has the potential to enhance single-cell analysis and facilitate its application across various research areas.

Highlights of scGPT include the following aspects:

Large-Scale Generative Foundation Model: scGPT is a large-scale generative foundation model pretrained on over 33 million cells, offering a unified framework for nonsequential omics data. Comprehensive experiments demonstrate the benefits of pretraining in both zero-shot and fine-tuning scenarios.

Alignment with Known Functional Groups: The gene networks learned by scGPT align strongly with known functional groups, highlighting the model's ability to capture biologically relevant information.

Pretraining universally and fine-tuning on demand: Through fine-tuning, the pretrained model's knowledge can be transferred to various downstream tasks such as cell type annotation, perturbation prediction, and multi-batch and multi-omic integration, enabling more accurate and biologically meaningful analyses.

4.3.1 Knowledge: Single-Cell Multi-omics

Single-cell multi-omics refers to the integration of multiple omics technologies at the single-cell level to study various molecular characteristics of individual cells simultaneously. Omics technologies include genomics, transcriptomics, epigenomics, proteomics, metabolomics, and others, each providing different layers of molecular information about cells. By combining these omics approaches, researchers can gain a more comprehensive understanding of the molecular mechanisms underlying cellular function and behavior.

Single-cell multi-omics can reveal intricate relationships between different molecular layers within individual cells, providing insights into how genetic information is translated into functional molecules like proteins and metabolites. This approach is particularly valuable for studying complex biological processes such as development, disease progression, and cellular response to external stimuli. By capturing multiple omics profiles from the same single cell, single-cell multi-omics can help uncover cellular heterogeneity, identify rare cell types, and elucidate regulatory networks and pathways at unprecedented resolution. The integration of multi-omics data also presents unique challenges, including data integration and computational analysis, which require sophisticated methods to extract meaningful insights from the complex datasets.

4.3.2 AI: Transformer, LLM, and Foundation Models

The Transformer is a deep learning model introduced by Vaswani et al. in 2017, primarily used for natural language processing tasks. It revolutionized the field by dispensing with recurrence (e.g., LSTMs) and instead relying on self-attention mechanisms to draw global dependencies between input and output. This architecture allows for parallelization and has become the foundation for many state-of-the-art models in NLP. In a Transformer, the input is tokenized into discrete units called tokens. Tokens are typically words, subwords, or characters, depending on the tokenization scheme used. Each token is then embedded into a high-dimensional vector space, which is learned during the training process.

GPT (Generative Pretrained Transformer) is a specific implementation of the Transformer architecture. GPTs, developed by OpenAI, are a series of models that use the Transformer architecture for various NLP tasks. The key innovation of GPT is its use of unsupervised pretraining on large text corpora, followed by fine-tuning on specific tasks. This pretraining allows GPT to learn rich representations of language, which can then be adapted to perform well on a wide range of downstream tasks such as language translation, text summarization, and question–answering.

A Foundation Model is a large, pretrained deep learning model that serves as a starting point for a wide range of tasks. These models are trained on massive datasets and learn to understand the nuances of language, making them capable of generating coherent text, answering questions, and performing other language-related tasks. Foundation models are designed to be versatile and can be fine-tuned on specific datasets or tasks to improve their performance in specialized areas. They serve as a foundational tool for developers and researchers, providing a powerful starting point for building NLP applications.

The self-supervised pretraining workflow in NLG, based on self-attention transformers, has shown promise in modeling input tokens of words. Applying a similar approach to single-cell data, where cells are characterized by genes and proteins, can enhance the understanding of cellular characteristics. The flexible nature of transformer input tokens allows for easy inclusion of additional features and meta-information. However, modeling genetic reads differs from natural language due to the nonsequential nature of gene order, i.e., no "next gene" corresponding to "next word" in text. Unlike words in a sentence, the order of genes is interchangeable, posing a challenge for applying causal masking formulations as used in GPT models. To address this, scGPT developed a specialized attention-masking mechanism that defines the order of prediction based on attention scores, tailored for modeling nonsequential omics data.

The core of scGPT consists of stacked transformer layers with multi-head attention, generating cell and gene embeddings simultaneously. The model undergoes two stages: first, general-purpose pretraining on 33 million human cells' scRNA-seq data under normal conditions, using a specially designed attention mask and generative training pipeline for self-supervised learning. This approach adapts scGPT to the NLG framework for sequential prediction. During training, the model learns to generate gene expression based on cell states or gene expression cues. The second stage involves fine-tuning on smaller datasets for specific applications, offering flexible pipelines for tasks like scRNA-seq integration, cell type annotation, and gene regulatory network inference in single-cell research.

References

1. Senior, A. W., Evans, R., Jumper, J., et al. (2020). Improved protein structure prediction using potentials from deep learning. *Nature, 577*(7792), 706–710.
2. Jumper, J., Evans, R., Pritzel, A., et al. (2021). Highly accurate protein structure prediction with AlphaFold. *Nature, 596*(7873), 583–589.
3. Alderson, T. R., Pritišanac, I., et al. (2023). Systematic identification of conditionally folded intrinsically disordered regions by AlphaFold2. *Proceedings of the National Academy of Sciences, 120*(44), e2304302120.
4. Barrio-Hernandez, I., Yeo, J., Jänes, J., et al. (2023). Clustering predicted structures at the scale of the known protein universe. *Nature, 622*(7983), 637–645.
5. Klukowski, P., Riek, R., & Güntert, P. (2023). Time-optimized protein NMR assignment with an integrative deep learning approach using AlphaFold and chemical shift prediction. *Science Advances, 9*(47), eadi9323.
6. Wayment-Steele, H. K., Ojoawo, A., Otten, R., et al. (2024). Predicting multiple conformations via sequence clustering and AlphaFold2. *Nature, 625*(7996), 832–839.
7. Abramson, J., Adler, J., Dunger, J., et al. (2024). Accurate structure prediction of biomolecular interactions with AlphaFold 3. *Nature, 630*, 493–500.
8. Watson, J. L., Juergens, D., Bennett, N. R., et al. (2023). De novo design of protein structure and function with RFdiffusion. *Nature, 620*(7976), 1089–1100.
9. Cui, H., Wang, C., Maan, H., et al. (2024). scGPT: Toward building a foundation model for single-cell multi-omics using generative AI. *Nature Methods 21*, 1470–1480.

Chapter 5
AI for Health and Medicine

5.1 Swarm Learning: Decentralized and Confidential Diagnosis

The goal of precision medicine is to identify patients with life-threatening diseases like leukemia, tuberculosis, or COVID-19. This is achieved by measuring molecular phenotypes using *omics* technologies and applying artificial intelligence (AI) approaches to use large-scale data for diagnostics. However, privacy legislation, like Convention 108+ of the Council of Europe, presents a challenge, limiting what is technically possible. Privacy-preserving AI solutions are crucial, especially during global crises, to answer important questions in fighting such threats. AI solutions depend on appropriate algorithms and large training datasets, but decentralization in medicine often results in insufficient local data for reliable classifiers. Centralization of data has been used to overcome this limitation, but it raises concerns about data ownership, privacy, and the creation of data monopolies. Federated AI, where data is kept locally, but model parameters are handled centrally, addresses some of these challenges, with risks such as power concentration and low fault tolerance.

Swarm Learning (SL) [1] is a decentralized learning system proposed to address the shortcomings of centralized data sharing in cross-institutional medical research. As shown in Fig. 5.1, SL combines edge computing, blockchain-based peer-to-peer networking, and coordination to maintain confidentiality without a central coordinator. Its blockchain technology provides robust measures against dishonest participants. SL enables confidentiality-preserving machine learning and can incorporate new developments in privacy algorithms. To demonstrate its feasibility, SL was used to develop disease classifiers for COVID-19, tuberculosis, leukemia, and lung pathologies using distributed data from over 16,400 blood transcriptomes and 95,000 chest X-ray images. SL outperformed classifiers developed at individual sites and complies with local confidentiality regulations.

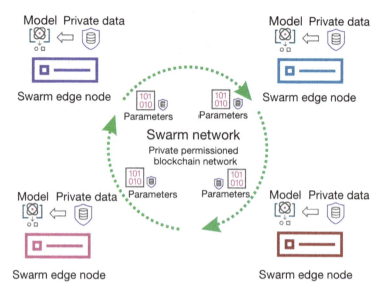

Fig. 5.1 Diagram of swarm learning for decentralized and confidential diagnosis. Full figure refers to Fig.1 in article *Swarm Learning for decentralized and confidential clinical machine learning* [1]

5.1.1 AI: Distributed Learning and Blockchain

Machine learning can be done locally with enough data and computer resources. In cloud computing, data is centralized for better results, but this leads to data duplication, increased traffic, and privacy challenges.

Distributed machine learning (DML) is a method of training machine learning models across multiple devices or servers, each processing a subset of the data. This approach allows for faster training of complex models by parallelizing the computation. DML is particularly useful when the dataset is too large to fit into the memory of a single machine or when training needs to be done on data located in different locations, such as in federated learning setups. In DML, the data is distributed across the devices or servers, and each device computes updates to the model based on its local data. These updates are then aggregated to produce a global model that benefits from the collective knowledge of all devices. DML requires coordination and communication between the devices to ensure that the global model converges to a good solution.

An alternative is Federated Leaning (FL), which combines multiple local leaners with private data. However, FL still uses a parameter server to aggregate local learning in a centralized manner.

Swarm Learning (SL) combines decentralized hardware, distributed machine learning, and permissioned blockchain for secure onboarding, leader election, and model parameter merging. SL uses an SL library (SLL) and decentralized data, avoiding a dedicated server and sharing parameters through the Swarm network. It ensures data sovereignty, security, and confidentiality through a private permissioned blockchain, with only pre-authorized participants able to execute transactions. New nodes join dynamically through blockchain smart contracts. SL divides into middleware and an application layer, with the application environment containing the machine learning platform, blockchain, and SLL, and the application layer containing the models.

5.1.2 Related Works on AI for Clinical Diagnostics

Digital pathology enables large-scale histopathological image analysis using deep learning. However, complexities in configuring the computational environment and optimizing hyperparameters limit deep learning's applications in this field. HEAL [2] is proposed as an automated framework for flexible and multifaceted histopathological image analysis. It is demonstrated in case studies on lung and colon cancer, offering an end-to-end tool for complex histopathological analysis and broadening deep learning's applicability in cancer image analysis.

Quantifying the pathogenicity of protein variants in disease-related genes is crucial for clinical decisions, yet over 98% of these variants have unknown effects. Computational methods could aid in interpreting these variants, but current approaches relying on labeled data are limited by sparse and biased labels. A new approach, EVE (evolutionary model of variant effect) [3], uses deep generative models to predict variant pathogenicity without labels. By modeling sequence variation across organisms, EVE captures constraints on protein sequences. It outperforms labeled data methods and matches or surpasses predictions from high-throughput experiments. EVE predicts pathogenicity for millions of variants and provides evidence for the classification of thousands of variants of unknown significance, offering valuable evidence for variant interpretation in research and clinical settings.

IRENE [4] is a transformer-based model for clinical diagnosis, designed to handle multimodal inputs like chief complaints, medical images, text, and structured data. Unlike other models, IRENE uses embedding layers to convert different types of data into unified visual and text tokens. With intramodal and intermodal attention, it learns holistic representations of patient data, outperforming other models in identifying pulmonary disease and predicting adverse outcomes in COVID-19 patients. This unified approach could streamline patient triaging and improve clinical decision-making.

5.2 Geneformer: Predicting Candidate Therapeutic Targets

Mapping gene regulatory networks that drive disease progression helps identify molecules that normalize core regulatory elements, rather than peripheral downstream effectors, potentially modifying diseases. However, mapping requires extensive transcriptomic data, hindering drug discovery in rare diseases or inaccessible tissues. Advances in sequencing have increased available data, and single-cell technologies offer precise, cell-specific data for network inference, crucial for diseases involving multiple cell types. Transfer learning, popular in natural language and computer vision, can leverage pretrained models for downstream tasks with limited data, democratizing knowledge across applications. Self-attention mechanisms in deep learning, like transformers, are adept at context-specific modeling, ideal for gene regulatory networks. Combining these methods, a new model Geneformer [5] uses attention-based deep learning pretrained on transcriptomic data to predict in data-limited settings, aiding discovery of network regulators and therapeutic targets.

5.2.1 Knowledge: Gene Regulatory Networks

Gene regulatory networks (GRNs) are systems of genes and regulatory elements (such as transcription factors, microRNAs, and epigenetic marks) that interact with each other and control the gene expression levels of mRNA or proteins in a cell. These networks play a crucial role in determining the cell's behavior, development, and response to external stimuli. Understanding GRNs can help researchers uncover the molecular mechanisms underlying various biological processes and diseases.

Mapping the gene network architecture requires large amounts of transcriptomic data because genes do not act in isolation; their expression levels are influenced by complex interactions with other genes and regulatory elements. To accurately infer these interactions and the overall network structure, researchers need data from a wide range of conditions, cell types, and states. This allows for the identification of patterns and correlations that reveal how genes regulate each other and how these interactions change in different biological contexts.

5.2.2 AI: Transformer for Gene

The pipeline of Geneformer has three steps: The first is an assembled dataset Genecorpus-30M with 29.9M human single-cell transcriptomes. The second one is pretraining of Geneformer on Genecorpus-30M using self-supervised masked learning, accurately predicting disease genes and targets. The third is fine-tuning Geneformer on downstream diverse tasks, boosting predictive accuracy, e.g., Geneformer identified therapeutic targets for cardiomyopathy, improving cardiomyocyte contraction in iPSC-based disease models.

5.2 Geneformer: Predicting Candidate Therapeutic Targets

Geneformer comprises six transformer encoder units, each with a self-attention layer and a feed-forward neural network layer. It has an input size of 2,048, embedding dimension of 256, four attention heads per layer, and a feed-forward size of 512. The model uses full dense self-attention across the input size. For gene embeddings, Geneformer maps each gene to a 256-dimensional space. Cell embeddings, representing the state of a single cell, are generated by averaging the gene embeddings detected in that cell, resulting in a 256-dimensional embedding.

Geneformer's pretraining is self-supervised, which means it can learn from unlabeled data. As an "attention-based" model, Geneformer learned to pay more attention to genes that play a key role in cells, through this self-supervised pretraining. Moreover, Geneformer also has context-aware capabilities, enabling it to make specific predictions based on the context of each cell. This is crucial because the function of genes is different crossing cell types, developmental stages, and disease states. The context awareness is particularly useful for studying diseases in which multiple cell types are affected and therapeutic targets may vary depending on the disease stage.

5.2.3 Data: Genecorpus-30M

Genecorpus-30M contains 29.9 million human single-cell transcriptomes from various tissues, excluding cells with high mutational burdens. It includes 561 datasets from droplet-based sequencing platforms, stored in the .loom HDF5 format for uniformity. Pretraining Geneformer with larger and more diverse datasets consistently improved its predictive power. Exposure to numerous experimental datasets during pretraining promoted robustness to technical artifacts and individual variability in single-cell analyses. These findings suggest that future models pretrained on even larger datasets may improve predictions in tasks with limited data.

5.2.4 PandaOmics: A Platform for Therapeutic Target Identification

PandaOmics [6] is a cloud-based platform that uses AI and bioinformatics to discover therapeutic targets and biomarkers from omics and text data. It generates hypotheses and provides evidence for these targets, validated in vitro and in vivo. It is part of Insilico Medicine's Pharma.ai suite, along with Chemistry42 and inClinico, for drug discovery. PandaOmics integrates omics and literature data into a Knowledge Graph (KG), extracting gene-disease associations and other insights. It also uses ChatPandaGPT, a large language model (LLM), for text summarization and contextualization. Integration with robotic platforms enhances target validation and compound screening, improving efficiency and accuracy. Robotic systems provide sequencing and phenotypic data, which PandaOmics uses to refine predictions, forming a feedback loop for target validation.

As the popular of ChatGPT, chatbot has been regarded as a new direction for therapeutic science. For example, perspectives including [7] and [8] discussed the potential application of chatbot.

5.3 drugAI: De Novo Drug Design Using Transformer with MCTS

Discovering new therapeutic compounds through de novo drug design is a significant challenge in pharmaceutical research. Traditional methods are resource-intensive and slow due to the vast molecular space. Computational methods like virtual screening and molecular dynamics have accelerated drug discovery but rely on existing molecules. Integrating AI and fostering collaboration can address biological complexity. Generative AI models are used in drug discovery but struggle to create entirely new structures. Self-supervised pretraining, similar to NLP, has been successful in training "chemical language models" on large datasets. These models treat chemical structures as sentences, with each symbol representing a chemical entity. To take the new advancements in AI, a novel de novo drug design engine drugAI [9] is introduced by integrating a decoder transformer model with MCTS, a first in bioinformatics and cheminformatics. The workflow of drugAI is illustrated in Fig. 5.2.

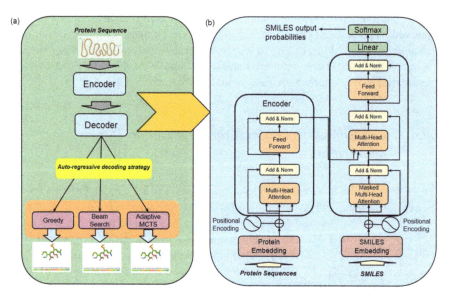

Fig. 5.2 The workflow of the translation task of drugAI. Full figure refers to Fig.3 in article *De Novo Drug Design Using Transformer-Based Machine Translation and Reinforcement Learning of an Adaptive Monte Carlo Tree Search* [9]

5.3 drugAI: De Novo Drug Design Using Transformer with MCTS

drugAI is an iterative approach that allows the model to improve its drug candidate generation, ensuring molecules meet physicochemical and biological constraints and bind effectively to targets. Results demonstrate drugAI's effectiveness across benchmark datasets, showing improved validity and drug-likeness compared to existing methods. Additionally, drugAI ensures generated molecules bind strongly to targets, highlighting its potential in accelerating drug discovery for various diseases.

5.3.1 SMILES: The Language for Drug Discovery

SMILES (Simplified Molecular Input Line Entry System) is a notation system used to represent the structure of chemical molecules in a simple and unambiguous way. It uses alphanumeric strings to represent atoms, bonds, and branches in a molecule. Each atom is represented by its atomic symbol (e.g., C for carbon and O for oxygen), and bonds between atoms are indicated by symbols such as "-" for a single bond and "=" for a double bond. Branches and ring structures are denoted by parentheses and numbers. SMILES strings can be used to encode molecular structures in databases, predict chemical properties, and conduct virtual screening for drug discovery.

5.3.2 AI: Transformer and MCTS

The core at drugAI is an encoder–decoder Transformer model with MCTS for drug discovery. drugAI is trained on protein–ligand pairs, filtered from the comprehensive BindingDB. The trained drugAI engine takes target protein sequences as input and generates small molecules (SMILES strings) as candidate inhibitors for these protein targets. The generation of molecules or sequences is in a step-by-step manner, i.e., the decoding, which involves adding amino acids or atoms one at a time to create new sequences or molecules, requires a decision-making strategy to select the best token at each step. The challenge with decoding algorithms like greedy search and beam search in drug design is ensuring that newly generated molecules meet certain constraints or properties necessary for effective drugs. Reinforcement learning (RL) with Monte Carlo Tree Search (MCTS) addresses this by introducing value functions, rewarding desired behaviors and penalizing undesired ones. In this way, drugAI ensures validity, binding to targets, and adherence to drug-likeness criteria.

5.3.3 Graph Neural Network in AI Pharmacy

Graph Neural Network (GNN) has drawn increasing attentions in recent years for pharmacy.

EmerGNN [10] is a graph neural network that can effectively predict interactions for emerging drugs by leveraging the rich information in biomedical networks. It extracts paths between drug pairs to learn pairwise representations, propagates information between drugs, and integrates relevant biomedical concepts. Weighted edges in the biomedical network indicate relevance for drug–drug interaction (DDI) prediction. EmerGNN outperforms existing methods in accuracy, identifying key information in the network for predicting interactions of emerging drugs.

Late-stage functionalization is a cost-effective method to enhance drug candidate properties. However, late-stage diversification is challenging due to chemical complexity of drug molecules. To overcome this, scientists created a platform using geometric deep learning and high-throughput reaction screening [11]. It employs two Graph Neural Network (GNN) architectures to predict reaction tasks, utilizing two FAIR datasets of 1,301 and 956 reactions. A new comprehensive reaction data format, SURF (a simple user-friendly reaction format), is developed to ensure FAIR data capture.

The study in [12] shows that deep graph learning models in drug discovery can be rendered explainable. They identified potential structural classes of antibiotics by using graph-based explanations of deep learning model predictions for antibiotic activity and cytotoxicity among 12,076,365 compounds. The study demonstrates that GNNs can be understood and explained more effectively by conducting graph-based searches for chemical substructure rationales that align with model predictions. This approach offers valuable chemical insights into the knowledge acquired by a specific model or ensemble of models.

References

1. Warnat-Herresthal, S., Schultze, H., Shastry, K. L., et al. (2021). Swarm learning for decentralized and confidential clinical machine learning. *Nature, 594*(7862), 265–270.
2. Wang, Y., Coudray, N., Zhao, Y., et al. (2021). HEAL: An automated deep learning framework for cancer histopathology image analysis. *Bioinformatics, 37*(22), 4291–4295.
3. Frazer, J., Notin, P., Dias, M., et al. (2021). Swarm learning for decentralized and confidential clinical machine learning. *Nature, 599*(7883), 91–95.
4. Zhou, H.-Y., Yu, Y., Wang, C., et al. (2023). A transformer-based representation-learning model with unified processing of multimodal input for clinical diagnostics. *Nature Biomedical Engineering, 7*(6), 743–755.
5. Theodoris, C. V., Xiao, L., Chopra, A., et al. (2023). Transfer learning enables predictions in network biology. *Nature, 618*(7965), 616–624.
6. Kamya, P., Ozerov, I. V., Pun, F. W., et al. (2024). PandaOmics: An AI-driven platform for therapeutic target and biomarker discovery. *Journal of Chemical Information and Modeling, 64*, 3961–3969.
7. Sin, J. (2024). An AI chatbot for talking therapy referrals. *Nature Medicine, 30*(2), 350–351.

References

8. Graber-Stiehl, I. (2023). Is the world ready for ChatGPT therapists? *Nature, 617*(7959), 22–24.
9. Ang, D., Rakovski, C., & Atamian, H. S. (2024). De Novo drug design using transformer-based machine translation and reinforcement learning of an adaptive Monte Carlo tree search. *Pharmaceuticals, 17*(2), 161.
10. Zhang, Y., Yao, Q., Yue, L., et al. (2023). Emerging drug interaction prediction enabled by a flow-based graph neural network with biomedical network. *Nature Computational Science, 3*(12), 1023–1033.
11. Nippa, D. F., Atz, K., Hohler, R., et al. (2024). Enabling late-stage drug diversification by high-throughput experimentation with geometric deep learning. *Nature Chemistry, 16*(2), 239–248.
12. Wong, F., Zheng, E. J., Valeri, J. A., et al. (2024). Discovery of a structural class of antibiotics with explainable deep learning. *Nature, 626*(7997), 177–185.

Chapter 6
AI for Chemistry

6.1 AlphaFlow: Discovering Materials via Bayesian Active Learning

Integration of machine learning (ML) with automated experimentation in chemistry and materials science has led to the development of self-driving labs (SDLs), revolutionizing research in these fields. SDLs combine automated physical experimentation with digital data processing and algorithm-guided experiment selection, enabling rapid exploration of complex problems. While proof-of-concept SDLs exist, truly autonomous research is limited to well-studied, constrained parameter spaces. Colloidal atomic layer deposition (cALD) for precision synthesis of quantum dots (QDs) exemplifies such challenges due to its high-dimensional experimental space. Previous studies have used SDLs with retrosynthetic planning algorithms for small molecule synthesis, but these approaches rely heavily on physics-based models and literature data, limiting their applicability to novel or complex systems. To overcome barriers like dimensionality and data scarcity in complex chemistries, new approaches are needed. Reinforcement learning (RL) is emerging as a powerful ML approach for such dynamic systems, offering the ability to handle sequence-dependent processes and predict the effects of individual steps in multistep reactions.

AlphaGo, the program that defeated a professional Go player in 2016, showcased the potential of reinforcement learning (RL) in complex decision-making. RL allows systems to learn from trial and error, creating data-rich environments without prior knowledge. In chemistry, RL-based algorithms have been applied in silico for process synthesis and route discovery. Integrating RL with closed-loop experimentation can address data scarcity and reproducibility issues in literature. AlphaFlow [1], an RL-guided self-driving lab (SDL), demonstrates this integration by autonomously exploring and discovering multistep chemistries, surpassing conventional methods. As shown in Fig. 6.1, AlphaFlow's modular fluidic processing units enable real-time iterative learning, identifying optimal synthetic routes for high-complexity

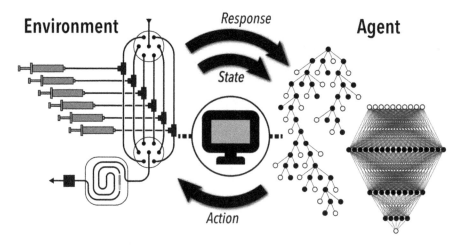

Fig. 6.1 Illustration of an RL-based feedback loop between the learning agent and the automated experimental environment. Full figure refers to Fig. 2 in article *AlphaFlow: autonomous discovery and optimization of multi-step chemistry using a self-driven fluidic lab guided by reinforcement learning* [1]

reactions. It navigates complex reaction spaces without prior knowledge, optimizing conditions for improved nanomaterial quality. This marks the first integration of RL with automated multistep chemistry, enabling rapid, intelligent, and constant exploration of complex reaction spaces, augmenting human researchers' capabilities.

6.1.1 AI: RL-Guided Self-driving Lab

As shown in Fig. 6.1, AlphaFlow is a self-driving lab (SDL) that integrates hardware and software to autonomously explore and optimize multistep chemical reactions.

The SDL hardware consists of modular fluidic microprocessors, which operate in a closed-loop system. The system starts with no prior information on the reaction sequence and rapidly generates data using reinforcement learning (RL) and a high-efficiency microdroplet flow reactor. Microscale flow reactors are used for their high efficiency and automation capabilities, enabling novel insights and precise control over reactive processes.

The SDL software utilizes RL-guided multistep synthesis. A single microdroplet reactor serves as the environment for the RL algorithm (agent). The agent evaluates the reactor's state and response based on prior actions and decides the next best action to navigate the high-dimensional reaction space intelligently and efficiently. The state is represented by a short-term memory (STM) containing the four prior injection conditions, accounting for relevant hidden parameters in the reaction space. The response is represented as a reward based on the in situ measured characteristics of the product. The agent includes a belief model composed of an ensemble neural network (ENN) regressor for reward prediction and a gradient-boosted decision tree for classifying state–action pairs. The belief model is constantly retrained on new experimental data to update its understanding and make informed decisions.

The belief model in AlphaFlow is constructed by training an ensemble neural network regressor and a gradient-boosted decision tree classifier on a fully formatted dataset. Each member of the regressor ensemble is trained with a randomly selected architecture and a random training set comprising 75% of the total data. The regressor maps state–action pairs to slope rewards, while the classifier maps state–action pairs to terminal or nonterminal conditions. The rollout policy uses the belief model to predict rewards for future action sequences, recommending the next action. It evaluates all possible action sequences up to four actions into the future, grouping them by their first action and applying a decision policy to determine the most valuable next action. During reaction space exploration, an upper confidence bound policy is used to maximize both the mean and standard deviation of predicted performance, aiming to explore regions with high potential and high model uncertainty. In exploitation experiments, the decision policy seeks only to maximize the mean predicted performance.

6.1.2 Knowledge: Colloidal Atomic Layer Deposition (cALD)

Colloidal atomic layer deposition (cALD) is a technique used in chemistry for the precise layer-by-layer growth of heteronanostructures at room temperature. It involves the sequential injection, removal, and washing of reactants and stabilizing ligands to grow these structures in a controlled manner. Compared to other shelling techniques, cALD offers monolayer precision, making it promising for synthesizing heteronanostructures with tailored properties. The self-limiting nature of cALD helps preserve the size dispersity of starting quantum dots (QDs) and allows for the tuning of confinement regimes and nanometer-scale heterostructure layers. Additionally, cALD can be applied to temperature-sensitive materials, such as metal halide perovskite QDs, due to its room temperature synthesis. However, cALD chemistry presents challenges, including an exponentially growing parameter space with each sequence step and the need for precise control over reaction sequence, concentrations, and time to prevent undesired reactions. Addressing these challenges requires innovative approaches beyond existing self-driving lab technologies.

6.1.3 Related Works

Automated chemical experiments by robots mark the beginning of AI labs. Despite existing AI-based systems, comprehensive scientific research remains challenging.

The authors [2] proposed AI-Chemist, with scientific data intelligence, performing basic chemical research tasks. It reads literature from a cloud database, proposes experiments, controls robots for synthesis, characterization, and testing, and analyzes data using machine learning. Three chemical tasks validated its competence, suggesting future AI-Chemists could transform chemical labs. The design-make test-analyze (DMTA) cycles for small molecule discovery are slow and laborious. By integrating machine learning (ML) generative algorithms, property prediction, computer-aided synthesis planning (CASP), robotics, and automated chemical processes, an autonomous chemical discovery platform [3] is developed. This platform can operate across various chemical spaces without manual reconfiguration. The study demonstrates an integrated DMTA cycle that uses prediction tools to propose, synthesize, and characterize molecules. In two case studies on dye-like molecules, 303 unreported molecules were discovered with desired properties.

Catalysts play an important role in chemical and chemical processes. Fast-Cat [4] quickly identifies the scalable performance of ligands for transition metal-catalyzed reactions with minimal ligand and time requirements. Its autonomous operation improves catalyst discovery by enhancing efficiency and speed, particularly in high-temperature/pressure and gas–liquid reactions. The algorithm's robustness generates large amounts of ligand-reaction data for modeling and optimization in ligand structure space, aiding in the search for novel ligand candidates. Fast-Cat's scalability bridges the gap between laboratory discovery and industrial reactor development. RoboChem [5], a robotic platform, streamlines photocatalytic reaction optimization, intensification, and scale-up. It combines accessible hardware, customized software, and a Bayesian Optimization (BO) algorithm to offer a hands-free, safe solution, eliminating the need for photocatalysis expertise.

6.2 Coscientist: Autonomous Chemical Research with Large Language Models

OpenAI released GPT-4 on March 14, 2023, demonstrating its application in chemistry-related problems amidst advances in automated chemical research. This progress includes autonomous discovery and optimization of organic reactions, automated flow systems, and mobile platforms. The integration of laboratory automation with powerful language models enables the development of a system that autonomously designs and executes scientific experiments. Scientists from CMU introduce Coscientist [6], an intelligent agent based on multiple language models, capable of autonomously designing, planning, and performing complex scientific experiments. Coscientist utilizes Internet browsing tools, robotic experimentation APIs, and other language models for tasks such as chemical synthesis planning, hardware documentation navigation, cloud laboratory command execution, precise liquid handling, complex scientific task execution, and optimization problem-solving using experimental data analysis.

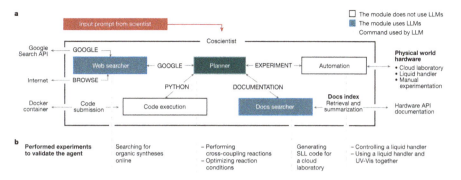

Fig. 6.2 Coscientist is composed of multiple modules that exchange messages. Full figure refers to Fig. 1 in article *Autonomous chemical research with large language models* [6]

6.2.1 AI: Generative Pretrained Transformer

Coscientist is an intelligent agent that uses multiple modules to solve complex problems, as shown in Fig. 6.2. The main module, called the "Planner," is powered by GPT-4 and acts as an assistant, planning based on user input using four defined commands: "GOOGLE" for web searching, "PYTHON" for calculations, "DOCUMENTATION" for accessing documentation, and "EXPERIMENT" for executing experiments through APIs. These commands interact with subactions to search the web, perform calculations, access documentation, and execute experiments using the Opentrons Python API and the Emerald Cloud Lab Symbolic Lab Language. The system protects user machines by using isolated Docker containers for code execution. The Planner's language model enables error correction in code and experiment execution. Overall, Coscientist demonstrates the ability to receive plain text input from users and autonomously perform tasks such as planning and executing experiments.

GPT (Generative Pretrained Transformer) is a type of language model developed by OpenAI. The GPT series are based on the Transformer architecture and are trained on large amounts of data to generate human-like text. They are "pretrained" in the sense that they are trained on a large dataset to learn the general structure and patterns of language before being fine-tuned for specific tasks. ChatGPT is a specific implementation or use case of the GPT model. It is a version of GPT that has been fine-tuned on conversational data, making it particularly well-suited for generating responses in a conversational context. ChatGPT is optimized for tasks like answering questions, engaging in dialogue, and providing information in a conversational manner.

GPT-4 is OpenAI's latest achievement in scaling up deep learning, representing a large multimodal model that accepts image and text inputs and produces text outputs. While not yet matching human performance in all scenarios, GPT-4 shows human-level performance on various professional and academic benchmarks. The model has been iteratively improved over 6 months, incorporating lessons from adversarial testing and ChatGPT to enhance factuality, steerability, and adherence to boundaries.

6.2.2 More Related Works Based on LLMs

In addition to Coscientist, there are methods and frameworks based on large-scale language models emerging in recent years.

ChemCrow [7] is a similar framework as the Coscientist. Large language models (LLMs) excel in various tasks but lack access to external knowledge sources, especially in chemistry. To address these issues, ChemCrow is designed as an LLM chemistry agent, integrating 18 expert-designed tools to enhance LLM performance in organic synthesis, drug discovery, and other chemical tasks. ChemCrow autonomously planned and executed syntheses and guided discovery, demonstrating effectiveness in diverse chemical tasks. ChemCrow not only aids chemists but also bridges experimental and computational chemistry, fostering scientific advancement.

In [8], large language models are leveraged for predictive chemistry. It shows that GPT-3 can be easily fine-tuned for chemistry tasks to answer chemical questions in natural language accurately. This approach performs well, especially with small datasets, even outperforming conventional techniques. The model's versatility suggests it could become a standard tool for bootstrapping projects and providing baselines for predictive tasks in these fields, leveraging its collective knowledge encoded in foundation models.

References

1. Volk, A. A., Epps, R, W., Yonemoto, D. T., et al. (2023). AlphaFlow: Autonomous discovery and optimization of multi-step chemistry using a self-driven fluidic lab guided by reinforcement learning. *Nature Communications, 14*(1), 1043.
2. Zhu, Q., Zhang, F., Huang, Y., et al. (2022). An all-round AI-chemist with a scientific mind. *National Science Review, 9*(10), nwac190.
3. Koscher, B. A., Canty, R. B., McDonald, M. A., et al. (2023). Autonomous, multiproperty-driven molecular discovery: From predictions to measurements and back. *Science, 382*(6677), eadi1407.
4. Bennett, J. A., Orouji, N., Khan, M., et al. (2024). Autonomous reaction Pareto-front mapping with a self-driving catalysis laboratory. *Nature Chemical Engineering, 1*, 240–250.
5. Slattery, A., Wen, Z., Tenblad, P., et al. (2024). Automated self-optimization, intensification, and scale-up of photocatalysis in flow. *Science, 383*(6681), eadj1817.
6. Boiko, D. A., MacKnight, R., Kline, B., et al. (2023). Autonomous chemical research with large language models. *Nature, 624*(7992), 570–578.
7. Andres M. B., Sam C., Oliver S., et al. (2023). ChemCrow: Augmenting large-language models with chemistry tools. arXiv, 2304.05376.
8. Jablonka, K. M., Schwaller, P., Ortega-Guerrero, A., et al. (2024). Leveraging large language models for predictive chemistry. *Nature Machine Intelligence, 6*(2), 161–169.

Chapter 7
AI for Material Science

7.1 CAMEO: Discovering Materials via Bayesian Active Learning

The discovery of new materials is one of the driving forces behind the advancement of modern science and technological innovation. However, traditional materials research and development involve extensive experiments, leading to low efficiency and high costs. To search for new materials, researchers not only need to conduct extensive theoretical investigations but also spend a considerable amount of time conducting experiments. For instance, if a researcher wants to analyze the characteristics of a material at N different temperatures, they would need to perform N experiments. Moreover, if there are M indicators to analyze in an experiment, each with ten values, the researcher would have to conduct 10^M experiments. With such a large number of experiments, researchers might spend several years or even decades. To efficiently conduct research and development on new materials, scientists from multiple institutions, including the National Institute of Standards and Technology (NIST) and the University of Maryland, developed an AI algorithm called CAMEO [1]. This algorithm effectively reduces the "repetitive" experimental time spent by scientists in the laboratory while maximizing research efficiency.

This real-time, closed-loop, autonomous system for materials exploration and optimization (CAMEO) is presented with several highlights:

CAMEO predicts the structure and functional properties of unknown materials by employing Bayesian machine learning to analyze data and utilizes active learning to determine the most valuable materials for further investigation.

CAMEO integrates scientists' experimental experience and domain knowledge gained from past experiments into the experimental workflow.

CAMEO discovers a novel epitaxial nanocomposite phase-change memory material, while reducing experimental time by tenfold.

CAMEO provides scientists with the capability for remote work.

7.1.1 Knowledge: Structural Phase Map

In the design of functional and structural materials, structural phase maps play a crucial role by outlining the relationship between materials' structure and composition. Traditionally, the generation of structural phase maps relied on expert knowledge and intuition involving an iterative process including materials synthesis, diffraction-based structure characterization, and crystallographic refinement, often spanning extended periods.

CAMEO revolutionizes this process through a materials-specific active-learning campaign. It pursues the dual objectives of maximizing knowledge of the phase map and identifying materials corresponding to property extrema. The subsequent phase mapping measurements in CAMEO are propelled by Bayesian graph-based predictions and risk minimization-based decision-making, ensuring that each measurement maximizes phase map knowledge. Acceleration of these tasks is achieved by leveraging mutual information via function integration, further boosted by incorporating physics knowledge (e.g., Gibbs phase rule) and prior experimental and theory-based insights into the target material system.

7.1.2 AI: Active Learning

Active Learning in AI is a machine learning paradigm where the algorithm actively selects specific data points from a larger dataset for labeling, with the goal of improving the model's performance while minimizing the amount of labeled data needed. In active learning, the algorithm intelligently chooses the most informative instances to query for additional labels, rather than relying on a randomly selected fixed set of labeled examples. The key idea is to strategically choose instances that are expected to be most beneficial for refining the model's understanding, reducing the need for extensive labeled datasets. This is particularly useful in situations where labeling data is expensive, time-consuming, or requires domain expertise.

The active learning process typically begins with the initialization, where the algorithm commences with a small set of labeled examples. Subsequently, the model undergoes training using the available labeled data. Following this, the instance selection phase ensues, during which the algorithm strategically picks instances from the unlabeled data that are perceived as either uncertain or highly informative. These selected instances are then presented to an oracle, which can be a human annotator or domain expert, for labeling. Once labeled, the model is updated with the new data, and the entire process iterates as the algorithm refines its understanding with each cycle, as shown in Fig. 7.1.

While the concept of "active learning" as understood in modern machine learning may not directly apply to historical scientific discoveries, one could draw a loose analogy to Laplace's guided discoveries in celestial mechanics during the eighteenth century. Laplace actively engaged in observational astronomy, theoretical

7.1 CAMEO: Discovering Materials via Bayesian Active Learning

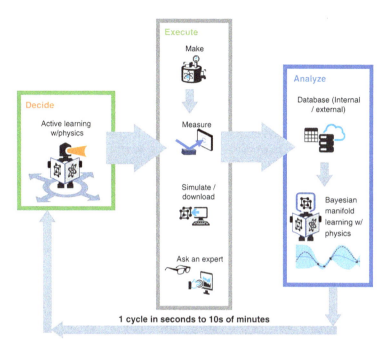

Fig. 7.1 Diagram of CAMEO. Figure refers to Fig. 1 in article *On-the-fly closed-loop materials discovery via Bayesian active learning* [1]

calculations, and mathematical modeling to understand the motion of celestial bodies. His keen insight and iterative approach, akin to active learning principles, involved refining hypotheses based on empirical observations and refining the models to better fit the observed data. Laplace's methodical and inquisitive approach to understanding celestial mechanics shares some parallels with the active learning framework, where the learner actively selects and utilizes information to enhance its understanding or model. While the terminology and formalism differ, the fundamental idea of actively seeking and utilizing relevant information for improved understanding connects the spirit of Laplace's scientific inquiries with the principles of active learning.

Bayesian Optimization (BO) is a probabilistic model-based optimization technique commonly used for optimizing complex, expensive, and noisy objective functions. This model is updated iteratively based on the observed data, and an acquisition function guides the selection of new points to evaluate in order to maximize the information gained about the objective function.

Active learning, in the context of Bayesian Optimization, involves selecting points in the input space strategically for evaluation. Instead of randomly selecting points, the algorithm actively chooses points that are expected to yield the most informative data to improve the model. This can lead to faster convergence and more efficient optimization, particularly in scenarios where evaluating the objective function is resource-intensive. The integration of Bayesian Optimization with active learning principles is essential to CAMEO, as it allows the algorithm to focus on the most relevant areas of the input space, reducing the number of expensive evaluations required to find an optimal solution.

For training data, CAMEO includes a database containing information about known stable phases, drawn from historical phase diagrams. This database serves as a reference for users to specify their material system of interest, with CAMEO autonomously identifying relevant phases. Additionally, two structural databases are integrated into the system. Structural data for these phases is automatically compiled using the Inorganic Crystal Structure Database (ICSD), which contains critically evaluated experimental structures, and the AFLOW.org density functional theory database.

7.1.3 Human Role

CAMEO brings the advantage of minimizing the consumption of valuable resources, including expert time, thereby allowing experts to focus on more advanced challenges. This is particularly noteworthy in research conducted at synchrotrons, where the demands and limitations on resources are exemplified.

A significant impact of CAMEO could be the empowerment of labs through AI control, potentially leading to a substantial reduction in the technical expertise required for experiment execution. This could herald a significant "democratization" of science, fostering a more distributed and inclusive approach to scientific endeavors, aligning with the principles of the materials collaboratory concept.

What aligns with the above is that CAMEO also illustrates an instance of human–machine interaction, involving human-in-the-loop participation in each cycle. In this scenario, humans contribute their expertise, while machine learning takes charge of decision-making steps. The real-time visualization of data analysis and decision-making, including the quantification of uncertainty, offers interpretability of the autonomous process for the human expert in the human–machine research team. CAMEO leverages the (currently) nonautomated skills of the human expert in the closed loop, thereby enhancing the capabilities of both human and machine.

7.2 GNoME: Graph Networks for Materials Exploration

In fields such as electric vehicle batteries, solar cells, and computer chips, the rapid discovery and application of new materials play a crucial role in driving industry development and innovation. Traditional materials research involves structural design through theoretical studies and reasonable assumptions, followed by attempts to synthesize materials, test their performance and characteristics, and continuously optimize and improve them. Therefore, the development of new materials is a slow

7.2 GNoME: Graph Networks for Materials Exploration

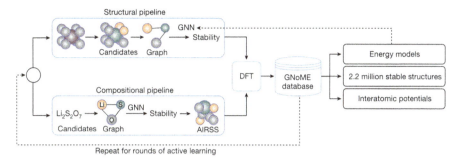

Fig. 7.2 Diagram of GNoME. Full figure refers to Fig. 1 in article *Scaling deep learning for materials discovery* [2]

and time-consuming process that requires extensive experimentation and validation over an extended period.

AI technology holds an advantage in the structural design and computational methods of new materials, accelerating the research process. However, it is essential that the discovered materials are "stable" and not easily decomposable to have significant implications for industrial applications. The Google DeepMind team proposed the Graph Networks for Materials Exploration (GNoME) [2], uncovering up to 2.2 million crystal structures that are theoretically stable but largely unexplored in experimental settings. Given structure or composition, graph neural networks (GNNs) are used to filter candidate structures, improving the efficiency of materials discovery by an order of magnitude, as shown in Fig. 7.2.

GNoME is highlighted with discovery of 2.2 million structures with high efficiency, building on 48,000 stable crystals identified in continuing studies. Of the stable structures, 736 have already been independently experimentally realized. GNoME also unlocks modeling capabilities for downstream applications, e.g., molecular dynamics simulations and prediction of ionic conductivity.

7.2.1 Density Functional Theory

Density Functional Theory (DFT) is a computational quantum mechanical modeling method used to study the electronic structure of many-body systems, especially atoms, molecules, and condensed matter. The primary goal of DFT is to calculate the electron density distribution rather than solving the Schrödinger equation for the many-body wave function, making it computationally more efficient for large systems compared to traditional quantum mechanical methods.

In DFT, the total electronic energy of a system is expressed as a functional of the electron density. The Hohenberg–Kohn theorems form the foundation of DFT, stating that the ground-state electron density uniquely determines the ground-state wave function and energy. The Kohn–Sham equations, named after Walter Kohn and Lu Jeu Sham, provide a practical implementation of DFT. These equations map the interacting system of electrons into a non-interacting system with an effective potential, making the calculations more tractable.

DFT has become a widely used tool in the fields of chemistry, physics, materials science, and biochemistry for predicting the properties and behavior of electronic systems. It allows researchers to study complex systems, such as large molecules or materials, providing valuable insights into electronic structure, chemical reactivity, and physical properties. Despite its success, DFT has limitations, and its accuracy depends on the chosen exchange–correlation functional, which approximates the unknown exchange–correlation energy. Advanced hybrid functionals and corrections have been developed to address some of these limitations and improve the accuracy of DFT calculations.

7.2.2 AI: Graph Neural Network

A Graph Neural Network (GNN) is a type of neural network designed to operate on graph-structured data. In traditional neural networks, data is typically represented as vectors or matrices. However, many real-world datasets, such as social networks, biological networks, and molecular structures, can be more naturally modeled as graphs. Graphs consist of nodes (representing entities) and edges (representing relationships or connections between entities).

GNNs are particularly well-suited for tasks involving graph-structured data. They can learn and analyze the intricate relationships and patterns within graphs, enabling them to make predictions or classifications based on the graph's topology and features associated with its nodes and edges. The key innovation of GNNs is their ability to perform operations on graph-structured data and update the node representations by considering both local and global information.

The architecture of a GNN typically involves a series of layers, each updating the node representations by aggregating information from neighboring nodes. This enables GNNs to capture complex dependencies and patterns within the graph, making them effective for tasks such as node classification, link prediction, and graph classification.

GNNs have found applications in various domains, including social network analysis, recommendation systems, bioinformatics, and materials science. They have proven to be powerful tools for understanding and making predictions in scenarios where the data can be naturally represented as a graph.

In the realm of materials science, crystal structures can be represented as graphs, wherein atoms serve as nodes, and edges depict interactions such as bonds. Graph Neural Networks are well-suited for managing such graph-structured data. In the realm of predicting the formation energy of crystal materials, GNNs prove valuable in modeling the intricate relationships and interactions among atoms within a crystal lattice. The GNN takes input encompassing information about the material's structural and compositional aspects. Through training, the network

7.2 GNoME: Graph Networks for Materials Exploration

learns to capture spatial arrangements, bond types, and other features influencing the material's formation energy. The adoption of GNNs for predicting formation energy brings forth advantages like leveraging graph representation, effectively capturing spatial information, adapting to diverse crystal structures, generalizing to new materials, and incorporating domain-specific knowledge for a comprehensive understanding of materials science.

In the context of structural models, the input comprises a crystal definition that encodes details such as the lattice, structure, and atom specifications. Each atom is denoted as an individual node within the graph, with edges established when the interatomic distance falls below a user-defined threshold. Node embedding is done based on atom type, while edge embedding is determined by the interatomic distance. On the other hand, for compositional models, the GNN processes input compositions by encoding them as a set of nodes. Each element type in the composition is represented by a node, and in compositional models, edges are created by default between all pairs of nodes in the graph.

Deep ensembles in machine learning refer to the use of multiple independently trained models, often neural networks, and combining their predictions to make more robust and accurate predictions. Each model in the ensemble is trained with different initializations or on different subsets of the training data, introducing diversity among the models. The final prediction is typically obtained by averaging or taking a vote over the predictions of individual models.

Utilizing an ensemble of models is a widely adopted and straightforward strategy to enhance the generalization and quantify predictive uncertainty in machine learning. This approach involves training n models instead of just one, where the prediction is determined by calculating the mean across the outputs of all n models, and the uncertainty is assessed by measuring the spread among the n outputs. In the context of stability prediction, GNoME employs $n = 10$ graph networks. Additionally, given the inherent instability in graph network predictions, the median is chosen as a more dependable performance predictor, and the interquartile range is utilized to define the range and bound uncertainty.

7.2.3 Dataset

GNoME endeavors to expand the repertoire of identified stable crystals, drawing inspiration from prior endeavors such as the Materials Project, OQMD, Wang, Botti, and Marques (WBM), and the ICSD. To ensure reproducibility, GNoME-based discoveries rely on static snapshots of two datasets captured at specific points in time, with the Materials Project data from March 2021 and the OQMD data from June 2021 serving as foundational references. For an updated evaluation, additional snapshots of the Materials Project, OQMD, and WBM were taken in July 2023. A total of approximately 216,000 DFT calculations were executed under consistent settings, enabling a comparative analysis of the rate of GNoME discoveries against those made by contemporaneous research initiatives.

7.3 A-Lab: An Autonomous Laboratory

The synthesis of novel materials is crucial for advancing technology across various industries, offering the potential to revolutionize electronics, healthcare, energy, and manufacturing. The creation of materials with unique properties enables breakthroughs in diverse applications, such as enhanced medical implants, improved energy storage solutions, and stronger lightweight materials.

However, material synthesis poses challenges, including the complexity of designing intricate molecular structures, the multidisciplinary nature requiring collaboration across scientific fields, the high costs and resource requirements, and the time-consuming process of experimentation and optimization. Additionally, concerns about environmental impact, scaling up from laboratory to commercial production, accurate characterization of material properties, and navigating regulatory frameworks add layers of complexity to the synthesis of novel materials.

Despite these challenges, researchers and industries continue to invest in overcoming these obstacles, recognizing the transformative potential of novel materials in addressing societal needs and advancing technological frontiers. Addressing the disparity in the pace between computational screening and the actual experimental synthesis of innovative materials, scientists present A-Lab [3]—an autonomous laboratory designed for the solid-state synthesis of inorganic powders, the principle of A-Lab is shown in Fig. 7.3.

A-Lab has a fully autonomous pipeline that integrates computations, historical data from the literature, active learning to plan and interpret the outcomes, natural language models to propose synthesis recipes, and experiments performed using robots. A-Lab has surprising achievement realizing 41 novel compounds over 17 days of continuous operation.

7.3.1 Solid-State Synthesis of Inorganic Powders

Discovering new materials through solid-state synthesis of inorganic powders necessitates a multifaceted expertise spanning several domains. Essential domains include (1) solid-state chemistry, covering principles like crystal structures, phase transitions, and inorganic thermodynamics, (2) materials science, involving knowledge of material properties and characterization techniques such as X-ray diffraction and scanning electron microscopy, and (3) inorganic chemistry for designing precursor materials, as well as proficiency in computational chemistry tools for simulating reactions and predicting material properties.

For a fully autonomous lab, researchers need familiarity with instrumentation, automation, and robotics for efficient experimental setups and proficiency in domain-specific software for predicting properties and analyzing experimental results. The integration of these diverse domains is crucial for successful materials discovery.

7.3 A-Lab: An Autonomous Laboratory

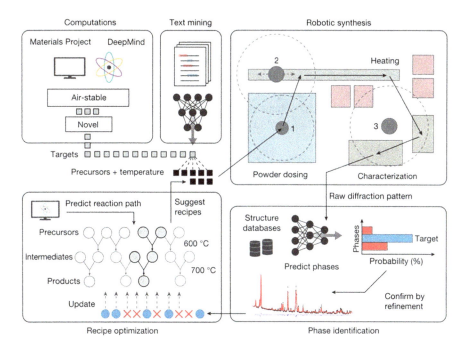

Fig. 7.3 Diagram of A-Lab. Full figure refers to Fig. 1 in article *An autonomous laboratory for the accelerated synthesis of novel materials* [3]

To be specific, the active learning runs under key domain knowledge (or assumptions) as follows:

Solid-state reactions tend to occur in pairs. In other words, a multiphase mixture will generally react two phases at a time. The most effective reaction pathway is the one that leads to maximal thermodynamic driving force at the target-forming step. Based on these two assumptions, A-Lab takes use of Autonomous Reaction Route Optimization with Solid-State Synthesis (ARROWS[3]) in the active-learning pipeline. ARROWS[3] prioritizes sets of precursors with large ΔG to form a user-specified target when starting a new experimental campaign. After the user has performed these experiments and fed their outcomes back to ARROWS[3], the package determines whether the attempted synthesis was successful, and if not, it learns which pairwise reactions formed detrimental intermediate phases that consumed the available free energy and therefore the target's formation. In subsequent experimental iterations, ARROWS[3] proposes new sets of precursors that it expects to avoid such intermediates and therefore maintain a larger thermodynamic driving force to form the desired target. The pairwise reactions learned by ARROWS[3] are stored in a local file that can be transferred between various experimental campaigns, enabling the algorithm to become more efficient as this reaction database grows.

7.3.2 AI: Active Learning with Robotics

This innovative platform integrates computational methodologies and historical data from literature and employs machine learning (ML) along with active learning to strategically plan and analyze experiment outcomes. The experimental procedures are executed using robotics, enhancing efficiency and bridging the gap between theoretical predictions and practical material synthesis.

Active Learning

Active learning is a machine learning paradigm that involves an algorithm actively selecting and acquiring new data points for training based on the current model's uncertainty or lack of confidence. Unlike traditional machine learning approaches that passively use fixed datasets for training, active learning allows the model to choose which instances to query for additional information, aiming to maximize the learning efficiency with minimal labeled data.

Active learning has proven valuable in scenarios where labeling large datasets is impractical, expensive, or time-consuming, making it a powerful approach to optimize the learning process in various machine learning applications.

The active learning cycle of the A-Lab aims to optimize the synthesis of novel inorganic materials. The active learning algorithm, implemented in ARROWS^3, is designed to automate the selection of optimal precursors for material synthesis by actively learning from experimental outcomes. Specifically, it determines which precursors lead to unfavorable reactions that form highly stable intermediates which prevents the formation of target material. Based on this, ARROWS^3 proposes new experiments using precursors it predicts to avoid such intermediates, thereby retaining a larger thermodynamic driving force to form the target. Experimental results show its ability to identify effective precursor sets for each target while requiring substantially fewer experimental iterations. This finding highlights the importance of domain knowledge in optimization algorithms for materials synthesis, which are critical for fully autonomous research platforms.

Robotics

There exists a gap between the rates of computational screening and the experimental realization of novel materials synthesis. While computational screening can be accelerated through high-performance computing, the experimental realization necessitates instrumental operations that involve significant human intervention. The low efficiency of experimental realization significantly slows down the procedure of discovering new materials.

A-Lab designed a fully automated pipeline by introducing robotic arms to handle and transfer samples between stations. The platform for experimental realization consists of five stations, i.e., a precursor preparation station for powder dispensing and mixing, a high-temperature heating station with four box furnaces, a product handling station developed in-house for powder retrieval and sample loading, and a characterization station with a powder X-ray diffractometer. The stations are independent, and three robot arms have been implemented to automate operations and connect the stations, forming a closed pipeline. More specifically, a central robot arm manages powder dispensing and mixing in the precursor preparation station, while two collaborative robot arms handle the transfer of samples and labware between stations.

7.3.3 Datasets

A-Lab takes use of several datasets for learning.

First, the large-scale ab initio phase stability data from the Materials Project and Google DeepMind is used to identify new desired target.

Second, a large database of syntheses extracted from the literature is used to train an ML model through natural language processing for generating initial synthesis recipes. The dataset is a knowledge base of 33,343 solid-state synthesis procedures extracted from 24,304 publications.

Third, a heating dataset from the literature is used to train a second ML model for proposing synthesis temperature.

Fourth, experimental structures from the Inorganic Crystal Structure Database (ICSD) are used to train probabilistic ML models for XRD patterns.

7.3.4 Human Role

Though A-Lab is designed to run autonomously, there are ways for scientists to control the pipeline. The operations of the lab are controlled through an application programming interface, which enables on-the-fly job submission from human researchers or decision-making agents.

References

1. Kusne, A. G., Yu, H., Wu, C., et al. (2020). On-the-fly closed-loop materials discovery via Bayesian active learning. *Nature Communications, 11*(1), 5966.
2. Merchant, A., Batzner, S., Schoenholz, S. S., et al. (2023). Scaling deep learning for materials discovery. *Nature, 624*, 80–85.
3. Szymanski, N. J., Rendy, B., Fei, Y., Kumar, R. E., et al. (2023). An autonomous laboratory for the accelerated synthesis of novel materials. *Nature, 624*, 86–91.

Chapter 8
AI for Astronomy

8.1 Locating Hidden Exoplanets

Exoplanet astronomy is the study of planets outside our solar system. This field is important because it helps us understand the prevalence, diversity, and formation of planets in the universe, as well as the potential for life beyond Earth. By studying exoplanets, astronomers can also gain insights into planetary systems' dynamics and evolution, providing valuable information for theories of planet formation and solar system evolution.

However, exoplanet astronomy faces several challenges. Detecting exoplanets is difficult due to their faintness compared to their host stars and their small angular separation. Characterizing exoplanets is also difficult, as it often requires advanced techniques to determine their atmospheric composition, surface conditions, and potential habitability. Additionally, the vast distances to exoplanetary systems make direct observation and detailed study challenging, requiring innovative approaches and technologies to overcome these limitations.

Recent studies have shown that analyzing the kinematics of protoplanetary disks is crucial for identifying unseen planets causing ring- and gap-like features. These features are due to the planets' spiral wakes, which create localized deviations from Keplerian motion known as "kinks." However, various mechanisms, such as gravitational instability, vertical shear instability, and magnetorotational instability, can also produce similar kinematic imprints. Traditionally, estimating the location of such planets involves computationally expensive hydrodynamics simulations and comparisons of synthetic and real observations by human. To address this problem, machine learning offers a more efficient approach. In astronomy, machine learning has seen significant growth, particularly in using computer vision techniques for analyzing protoplanetary disk images. Previous work has used deep learning to infer planet properties, but these efforts did not incorporate kinematic information.

Incorporating machine learning into kinematic observations could improve the accuracy and speed of identifying and characterizing planets within protoplanetary disks. To this end, scientists present a machine learning model, trained on synthetic data from SPH simulations, which accurately identifies forming exoplanets in telescope observations of protoplanetary disks [1]. When applied to HD 97048 and HD 163296 systems, the models predict planets with over 99% confidence. In HD 97048, the model's prediction aligns with previous studies, confirming the planet's presence and location.

8.1.1 ALMA Observations

ALMA stands for the Atacama Large Millimeter/submillimeter Array, which is a powerful astronomical observatory located in the Atacama Desert of northern Chile. ALMA consists of an array of 66 high-precision antennas that work together as a single telescope to observe light at millimeter and submillimeter wavelengths. These wavelengths are longer than those of visible light, allowing ALMA to see through the dense clouds of dust and gas where stars are born.

ALMA's unique capabilities make it ideal for studying a wide range of astronomical phenomena, including the formation of stars and planets, the chemistry of interstellar space, and the structure of galaxies. ALMA observations have provided valuable insights into the early universe, the formation of planetary systems, and the processes that drive star formation and galaxy evolution.

ALMA data are used by astronomers around the world to address fundamental questions about the universe. The observatory is a collaboration between East Asia, Europe, and North America, with contributions from other countries. Its location in the Atacama Desert, at an altitude of 5,000 m (16,400 feet), provides clear skies and minimal atmospheric interference, making it one of the best places on Earth for millimeter and submillimeter astronomy.

In addition to ALMA, which continues to deliver larger and larger disk survey datasets, the next-generation telescopes—such as JWST, ngVLA, and the Square Kilometre Array (SKA)—will increase the time and cost of analyzing the growing amount of data. AI methods show promise in addressing this challenge.

8.1.2 Simulation for Synthetic Data

The models are trained on synthetic images generated from simulations before applied to real observations. The simulation method is the 3D Smoothed Particle Hydrodynamics (SPH), which is a computational method used primarily for simulating fluid flows and other continuum mechanics problems. In SPH, the fluid is represented by a set of particles that move through the simulation domain. Each particle carries various properties such as density, velocity, and pressure. The properties of a particle are determined by averaging the values of neighboring particles within a smoothing kernel. This smoothing process allows SPH to model

8.1 Locating Hidden Exoplanets

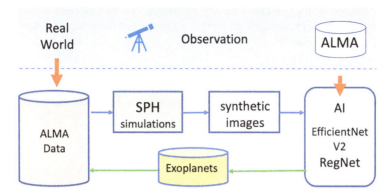

Fig. 8.1 Parallel learning framework for locating hidden exoplanets, using both simulated and real data

complex fluid behavior, including free surface flows, fluid mixing, and fluid–structure interactions. SPH is widely used in astrophysics, engineering, and other fields where fluid dynamics simulations are needed.

The experiment runs 1000 3D SPH Simulations; 25% were used for testing only. The simulation results are used to create velocity channel maps, as shown in Fig. 8.1. Points between 10 and 100 orbits are randomly selected for each simulation. To address class imbalance, six snapshots are chosen from simulations without planets and two from simulations with planets. This creates a velocity cube with 600×600 pixel images for model input. The channels are convolved spatially and spectrally, and noise is added by Latin Hypercube sampling. Stacking all velocity channels into a single image helps address the challenge of kinks appearing in only a few channels, potentially neighboring the channel containing the planet.

8.1.3 EfficientNet and RegNet

The models take in a $(600 \times 600 \times C)$ image, and then the output is a two-component vector passing through Softmax to predict the probability that the input belongs to "contains at least one planet" class or "does not contain a planet" class.

Two models, EfficientNetV2 and RegNet, are employed, with EfficientNetV2 focusing on performance and parameter efficiency, while RegNet enhances performance through convolutional recurrent neural networks.

Trained on synthetic data, these models are then applied to real telescope data, specifically the HD 163296 and HD 97048 systems. They successfully replicate predictions and locations of forming exoplanets, with over 99% confidence in both cases. The prediction for HD 97048 aligns well with a previous study.

Although this work is a kind of proof-of-concept, it demonstrates that machine learning can automate the task and exceed human proficiency, even when high confidence is demanded.

8.2 DLPosterior: Estimating Dark Matter Distribution

Dark matter is a mysterious form of matter that makes up about 27% of the total mass and energy content of the universe. Dark matter is believed to be a fundamental component of the universe, alongside ordinary matter and dark energy. Determining its nature is crucial for understanding the overall composition and evolution of the cosmos. Dark matter plays a crucial role in the formation and evolution of large-scale cosmic structures. The nature of dark matter could provide insights into fundamental physics beyond the Standard Model, which currently does not account for dark matter. Understanding dark matter is essential for understanding the formation and evolution of galaxies, as dark matter halos are thought to provide the gravitational scaffolding around which galaxies form.

However, unlike ordinary matter, which consists of atoms and emits, absorbs, or reflects light, dark matter does not interact with electromagnetic radiation and thus cannot be seen directly. Therefore, the presence of dark matter can only be inferred from its gravitational effects on visible matter, light, and the large-scale structure of the universe.

Weak lensing mass-mapping is one of the key methods of reconstructing matter distribution maps from measured galaxy ellipticities, but it faces challenges due to noise and missing data, making the problem ill-posed. Various methods use different prior assumptions to regularize the problem, such as the Kaiser–Squires method, which directly inverts the lensing operator with Gaussian smoothing. Other approaches use maximum entropy, Gaussian prior, and sparsity regularization, but they often lack proper uncertainty quantification.

Bayesian methods aim to recover a posterior probability estimate but are constrained by restrictive prior assumptions. Recent advances involve deep neural networks, like DeepMass, which utilizes a U-net with a simulation-defined prior, and models based on generative adversarial networks (GANs), which denoise weak lensing mass-maps and consider the spherical curvature of the sky. However, Bayesian approaches require simplifications due to the intractable analytical distribution of mass-maps, which can only be sampled from.

To deal with these issues in mass-mapping, scientists introduce a novel method, DeepPosterior [2] by merging deep learning with Bayesian inference. It offers a practical means to sample from the complete high-dimensional posterior distribution of convergence maps. The technique entails deriving a prior from samples obtained from an implicit distribution (e.g., simulated convergence maps) and employing this prior to sample the entire Bayesian posterior. As shown in Fig. 8.2, the result of DeepPosterior is superior to that of previous methods.

8.2 DLPosterior: Estimating Dark Matter Distribution

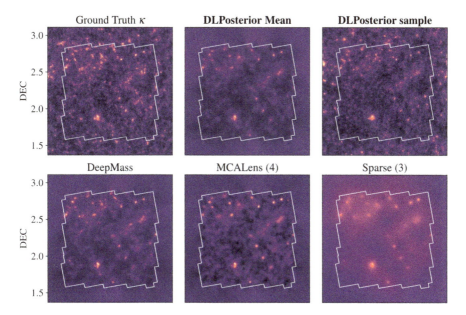

Fig. 8.2 Diagram of DLPosterior. Full figure refers to Fig. G1 in article of [2]

DLPosterior is distinguished in several aspects. DLPosterior is designed from a unified view of the mass-mapping problem, framing it as a Bayesian posterior estimation issue. The DLPosterior framework leverages numerical simulations to establish a comprehensive non-Gaussian prior for the convergence field. This approach enables sampling from the complete Bayesian posterior, incorporating both the simulation-driven prior and a physical likelihood. DLPosterior uses a Denoising Score Matching (DSM) technique to learn a prior from high-resolution hydrodynamical simulations and employs an annealed Hamiltonian Monte Carlo (HMC) approach for efficient sampling of the high-dimensional Bayesian posterior.

DLPosterior demonstrates quantitative improvement over standard methods, including Kaiser–Squires inversion, Wiener filter, or GLIMPSE2D. Furthermore, DLPosterior provides uncertainty quantification with the posterior samples, such as posterior variance, which is lacking in previous methods.

8.2.1 Mass-Mapping and Gravitational Lensing Effect

The Mass-Mapping Problem
The mass-mapping problem in astronomy refers to the challenge of mapping the distribution of dark matter in the universe using observations of gravitational lensing. Dark matter does not emit, absorb, or reflect light, so it cannot be directly

observed. However, its presence can be inferred from its gravitational effects on visible matter and light.

Gravitational lensing occurs when the gravitational field of a massive object, such as a galaxy or a cluster of galaxies, bends the path of light from more distant objects behind it. This bending can distort the images of these background objects, creating arcs, multiple images, or other effects. By analyzing these distortions, astronomers can infer the distribution of mass (both visible and dark) in the foreground object.

The mass-mapping problem involves using sophisticated modeling techniques to reconstruct the mass distribution of the foreground object (such as a galaxy cluster) from the observed gravitational lensing effects. This is challenging because gravitational lensing depends on both the distribution of mass and the geometry of the system, and the exact relationship between the observed distortions and the underlying mass distribution is complex. By solving the mass-mapping problem, astronomers can create detailed maps of the dark matter distribution in the universe, shedding light on its nature and role in cosmic structure formation.

Weak Gravitational Lensing Effect

Weak gravitational lensing is a subtle effect predicted by Einstein's theory of general relativity, where the gravitational field of a massive object (such as a galaxy or a cluster of galaxies) slightly distorts the light from more distant objects behind it. This distortion causes a slight stretching or shearing of the images of background galaxies, which can be observed as small changes in their shapes.

The weak gravitational lensing effect is called "weak" because the distortions it produces are typically small, on the order of a few percent or less. This is in contrast to strong gravitational lensing, where the distortions are much more pronounced and can lead to the formation of multiple images or arcs around the lensing object.

By measuring the statistical properties of the shapes of large numbers of background galaxies, astronomers can infer the presence and distribution of dark matter in the foreground objects causing the lensing. Weak gravitational lensing is a powerful tool for studying the large-scale distribution of dark matter in the universe, as it allows astronomers to map out the "invisible" dark matter that cannot be directly observed.

8.2.2 AI: Score-Based Generative Modeling

The DLPosterior framework integrates several AI methods, i.e., Denoising Score Matching, Score-Based Generative Modeling, and Annealed Hamiltonian Monte Carlo.

Denoising Score Matching (DSM)

Denoising Score Matching (DSM) is a technique used in machine learning and statistical modeling to estimate probability distributions from noisy data. It is particularly useful in scenarios where the true underlying distribution is unknown but can be inferred from noisy samples.

8.2 DLPosterior: Estimating Dark Matter Distribution

DSM involves two key steps: denoising and score matching. (1) Denoising: The first step involves removing noise from the observed data to obtain a cleaner representation of the underlying distribution. This is typically done using a denoising autoencoder or a similar method that learns to reconstruct the original data from noisy inputs. (2) Score Matching: Once the data is denoised, the next step is to estimate the score function of the underlying distribution. The score function provides information about the gradient of the log-likelihood function, which is crucial for estimating the distribution itself. Score matching involves minimizing the discrepancy between the estimated score function and the true score function of the underlying distribution.

In DSM, deep neural networks (DNNs) are utilized primarily in the denoising step of the process. Particularly, autoencoders are commonly employed to denoise the observed data. An autoencoder consists of an encoder network that compresses the input data into a lower dimensional representation (latent space) and a decoder network that reconstructs the original input from the latent representation.

During training, the autoencoder is trained to minimize the reconstruction error, effectively learning to remove noise from the input data. The denoising autoencoder is trained on noisy samples generated by corrupting the clean data with various forms of noise. This process encourages the autoencoder to learn a robust representation of the underlying structure in the data, making it effective at denoising.

Once trained, the denoising autoencoder can be used to remove noise from observed data samples. These denoised samples can then be used in subsequent steps of the DSM algorithm, such as estimating the score function or learning the underlying probability distribution.

By incorporating deep neural networks in the denoising step, DSM can effectively handle noisy data and estimate the underlying probability distribution more accurately.

Score-Based Generative Modeling
Score-based generative modeling is a class of generative modeling approaches that directly estimate the score function of a probability distribution, rather than explicitly modeling the distribution itself. The score function provides information about the gradient of the log-likelihood function, which is crucial for sampling from and learning the underlying distribution.

In score-based generative modeling, a neural network is typically used to approximate the score function. This neural network is trained to output the gradient of the log-likelihood with respect to the input data. Once the network is trained, it can be used to generate samples from the distribution by performing a gradient ascent or descent procedure from a given starting point.

Score-based generative modeling has several advantages, including the ability to model complex and high-dimensional distributions, the ability to capture multimodal distributions, and the potential for more stable training compared to other generative modeling approaches like GANs or VAEs. However, it is computationally expensive, as it requires estimating the score function at each step of the sampling process.

Annealed Hamiltonian Monte Carlo
Annealed Hamiltonian Monte Carlo (HMC) is a variant of the Hamiltonian Monte Carlo algorithm, a powerful method used for sampling from complex probability distributions. HMC is particularly effective in high-dimensional spaces where other methods might struggle, such as in Bayesian inference problems.

The "annealed" part in Annealed HMC refers to the use of an annealing schedule, where the algorithm gradually transitions from sampling a simpler, approximate distribution to the target distribution of interest. This gradual transition helps the sampler explore the target distribution more effectively, especially when the target distribution is complex or has multiple modes.

In the context of HMC, the annealing schedule is typically applied to the mass matrix or the step size used in the algorithm. By starting with a simpler distribution (e.g., a Gaussian distribution) and gradually increasing the complexity towards the target distribution, Annealed HMC can overcome potential issues like getting stuck in local modes and improve the overall sampling efficiency.

Annealed HMC is a valuable technique for sampling from complex distributions, especially in situations where traditional sampling methods may struggle to explore the space effectively.

8.3 Astroconformer: Analyzing Stellar Light Curves

In astronomy, a light curve is a graph that shows the brightness of an object (such as a star, a variable star, or a supernova) as a function of time. Light curves are used to study the variability of celestial objects, including changes in brightness caused by eclipses, pulsations, explosions, or other phenomena. By analyzing light curves, astronomers can learn about the nature, evolution, and physical processes occurring in these objects.

Traditional asteroseismic analyses use power spectra from light curves to estimate oscillation properties but face challenges in dwarf stars and need high-cadence observations for main-sequence stars.

To address these challenges, scientists proposed a new Transformer-based deep learning model called Astroconformer [3]. As shown in Fig. 8.3, Transformers are adept at capturing long-range correlations, making them ideal for analyzing time series data like stellar light curves. This approach allows for direct analysis of observed light curves in the time domain, minimizing information loss and eliminating the need for additional postprocessing steps.

Astroconformer outperforms k-nearest neighbor-based methods and convolutional neural networks in inferring log g (gravity) values. Astroconformer achieves a relative median absolute error of 1.93% and a relative median standard deviation of 2.10% in predicting v_{max} (frequency of maximal oscillation power) from Kepler

8.3 Astroconformer: Analyzing Stellar Light Curves

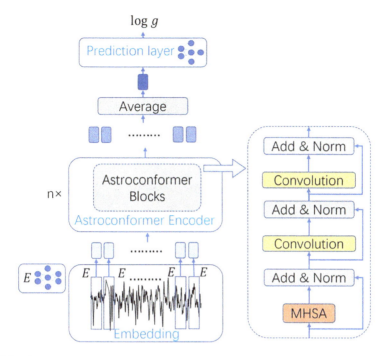

Fig. 8.3 Diagram of Astroconformer. Full figure refers to Fig. 3 in article of [3]

light curves, outperforming the SYD pipeline. For shorter 33-day Kepler light curves, Astroconformer achieves a relative median absolute error of less than 3% and reliably estimates v_{max}, unlike the conventional pipeline, which fails in approximately 30% of cases. Astroconformer's attention maps demonstrate its ability to identify stellar oscillations and longer period granulation patterns, enhancing log g estimations, particularly for less evolved stars.

8.3.1 Stellar Light Curves

Stellar light curves, shaped by factors like rotation, spot modulation, oscillations, and granulation, are crucial for understanding stellar properties.

Asteroseismology, studying stellar oscillations, is key for determining mass, radius, and luminosity, especially in evolved stars. Traditionally, asteroseismic analyses transform light curves into power spectra, focusing on parameters like v_{max} and Δv to estimate oscillation properties. However, this approach has limitations, such as challenges in detecting oscillations in dwarf stars and the need for high-cadence observations for main-sequence stars. Moreover, not all giants exhibit solar-like oscillations. The upcoming Rubin Observatory and Zwicky Transient

Facility have long cadences that may hinder oscillation detection. Therefore, innovative methodologies are needed for robust asteroseismic analysis.

Stellar granulation, correlated with mass and evolutionary stage, offers an alternative for characterizing stars using long-cadence data. Granulation, covering a broader timescale range than oscillations, is suggested as a more effective method for subgiants and dwarfs. Overall, stellar light curves contain valuable information beyond oscillations, offering insights into additional stellar characteristics if fully utilized.

8.3.2 AI: Self-attention and Transformer

Asteroseismology has increasingly utilized machine learning (ML) to extract insights from stellar light curves beyond oscillation data. Techniques like polynomial ridge regression and k-Nearest Neighbors (k-NNs) have been applied to light curve autocorrelation functions and power spectra, respectively, achieving precise log g estimates for red giants. Convolutional Neural Networks (CNNs) and Recurrent Neural Networks (RNNs) have also been used for image recognition and sequence modeling in asteroseismology. However, these ML methods have limitations, such as k-NN's challenges with generalization and CNNs' and RNNs' difficulties in capturing long-range correlations. New advances in Transformer provide a more powerful approach.

Multi-head Self-attention

Sequence representation is a key focus in deep learning, especially for tasks like NLP, CV, and time-series forecasting. Unlike images, sequences require capturing long-range correlations. Self-attention, a crucial concept, analyzes correlations between timestamps and adjusts contributions from each timestamp. It duplicates the input sequence and projects it into query, key, and value forms, using learnable parameters. The query seeks relevant information, the key aligns with the query, and the value decides how much original information to include in the output. This process creates a new sequence representation based on attention mechanisms. However, single-head self-attention, which computes attention using dot products, has limitations in expressiveness. To address this, Multi-Head Self-Attention (MHSA) extends the concept, allowing the model to capture various types of correlations within the sequence.

Astroconformer combines convolutional layers for local information and MHSA layers for long-range correlations in analyzing stellar light curves. It uses an approach that divides light curves into fixed-size patches, each containing 20 timestamps, and projects them into a 128-dimensional space using an MLP. This allows Astroconformer to extract both short-range features and cross-correlate features across different time spans, enhancing its ability to analyze light curves effectively. Despite variations in light curve length, Astroconformer consistently segments them into 20-timestamp patches for analysis.

The Astroconformer encoder consists of multiple blocks, each containing an 8-head MHSA layer and two convolutional modules. It omits feed-forward modules, focusing on parameter-efficient modules that enable meaningful interactions between patches. To address MHSA's agnosticism to sequence order, Rotary Positional Encoding (RoPE) is used in each MHSA module.

The Embedding layer and Astroconformer encoder produce a 200 × 128 patch embedding sequence from the input light curve. Average pooling along the temporal dimension results in a final 128-dimensional embedding of the light curve. This approach allows Astroconformer to handle light curves of varying lengths. The 128-dimensional final embedding then undergoes processing through a two-layer MLP to predict $\log g$.

Dataset

Asteroseismology has advanced significantly with missions like CoRoT and Kepler, which provided high-quality light curves for hundreds of thousands of stars. Kepler's original mission lasted 4 years, focusing on low- to intermediate-mass stars. The K2 mission extended this with observations of various ecliptic fields, though with shorter observation windows. TESS, Kepler's successor, monitors bright stars across the sky but poses challenges with shorter observation periods, ranging from 27 days to about a year.

References

1. Terry, J. P., Hall, C., Abreau, S., et al. (2022). Locating hidden exoplanets in ALMA data using machine learning. *The Astrophysical Journal, 941*(2), 192.
2. Remy, B., Lanusse, F., Jeffrey, N., et al. (2023). Probabilistic mass-mapping with neural score estimation. *Astronomy & Astrophysics, 672*, A51.
3. Pan, J.-S., Ting, Y-S., Yu, J. (2024). Astroconformer: The prospects of analyzing stellar light curves with transformer-based deep learning models. *Monthly Notices of the Royal Astronomical Society, 528*(4), 5890–5903.

Chapter 9
Toward a Sustainable AI4S Ecosystem

9.1 A Brief Summary of AI4S from Viewpoint of HANOI

AI Methods

Recent advancements have seen various AI models and algorithms tackling complex problems across different fields. Convolutional Neural Networks are used in astronomy for tasks like galaxy classification and object detection in sky surveys. Graph Neural Networks are applied in biology to predict protein–protein interactions and in materials science to predict material properties based on atomic structures. In chemistry, Reinforcement Learning is used for drug discovery, where agents generate novel molecular structures with desired properties. Transformer models are applied in biology for analyzing genetic sequences and in physics for simulating complex physical systems. Additionally, Bayesian Optimization is used in experimental design and hyperparameter tuning, optimizing resource allocation and parameter settings in scientific experiments.

In addition, large language models (LLMs) have been increasingly used in AI for sciences. LLMs, such as GPT-3, have been applied in various scientific fields for tasks such as text generation, summarization, and even problem-solving. For example, LLMs summarize scientific papers, extracting key information and reducing the time researchers spend reading; analyze and interpret complex scientific data, helping researchers gain insights more efficiently; and assist in solving complex scientific problems by providing suggestions, hypotheses, or explanations based on existing knowledge.

Domain Knowledge

Domain knowledge plays a crucial role in AI for sciences, enhancing the understanding and performance of AI models.

For example, in physics, domain knowledge helps in defining physical laws and constraints, guiding the development of accurate simulations and models.

Knowledge of fundamental particles and their interactions informs the design of AI models for data analysis at facilities like the Large Hadron Collider.

In chemistry, domain knowledge helps in defining molecular structures and properties, guiding the development of AI models for drug discovery and materials science. For instance, understanding chemical bonding and reactivity aids in the design of AI algorithms for predicting molecular properties and reactions.

In astronomy, domain knowledge about celestial objects and phenomena guides the development of AI models for data analysis and image processing. For example, knowledge of stellar evolution informs the design of AI algorithms for classifying stars based on their spectral features.

As we have mentioned in Chap. 1, the fusion of AI methods and domain knowledge, as well as collaborations between scientists in different domains, are essential for successful AI4S applications.

Virtual System and Simulations

Simulations play a crucial role in AI for sciences by providing data for training machine learning models, validating algorithms, and exploring complex systems. We have seen many examples across different disciplines:

In Physics, simulations in physics help understand phenomena like particle interactions in high-energy physics (e.g., Large Hadron Collider simulations), cosmological simulations of the universe's evolution, and simulations of quantum systems.

In chemistry, molecular dynamics simulations are used to study the behavior of molecules and reactions, aiding drug discovery (e.g., predicting protein–ligand interactions) and material science (e.g., studying properties of materials at atomic scale).

In biology, simulations are used to model biological systems, such as protein folding (e.g., Folding@home project), gene regulation networks, and cellular processes, aiding in understanding diseases and drug development.

In astronomy, simulations help model the formation and evolution of galaxies, stars, and planetary systems, providing insights into cosmic phenomena.

Data

We have also seen that data, both real and synthetic, play a crucial role in AI4S by providing the basis for training and validating machine learning models. Disciplines including physics, chemistry, biology, astronomy, etc. have benefit from synthetic data generated from simulations. The real data provides insights into natural phenomena, while synthetic data augments real data to train models, validate predictions, and explore scenarios that may be challenging or impossible to observe in the real world.

Roles of Scientists

The roles of scientists in the research process have changed with AI to help in data analysis, hypothesis generation, experimental design, scientific literature review, decision support, and more. AI for Sciences has the potential to revolutionize scientific research by accelerating discovery, improving accuracy, and enabling scientists to tackle complex problems more effectively.

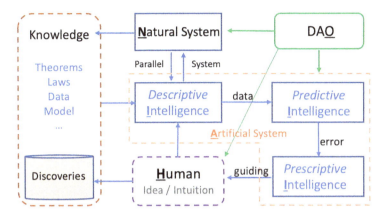

Fig. 9.1 HANOI-AI4S: a general framework for AI4S based on Parallel Intelligence

Nearly all of the AI4S works in preceding chapters have covered the factors above. However, from viewpoint of Parallel Intelligence, there is a key factor that has been largely overlooked. As shown in the upper right of Fig. 9.1, this overlooked factor is organization (e.g., DeSci and DAO), which will be the corner stone of AI4S to organize all resources to build a vibrant and sustainable ecosystem.

9.2 Toward AI4S Ecosystem

9.2.1 The Origins and Goals of AI4S

AI for Science has made significant strides across various scientific disciplines, including physics, chemistry, biology, astronomy, and so on. With the AI advancements such as Large Language Models (LLMs), AI4S continues to advance, offering new methods to tackle complex scientific problems and accelerating the pace of discovery.

The scientific researchers should also look from the reverse side of S4AI (Science for AI), especially SS4AI (Social Science for AI), as shown in Fig. 9.2, the core of which is the ethics and governance issues of artificial intelligence and broader intelligent science and technology. It must be realized that from AlphaGo to ChatGPT, the current cutting-edge artificial intelligence technology cannot be explained, and intelligence in a broad sense cannot be scientifically explained in its connotation; though artificial intelligence cannot be explained, it must be able to be governed, and this is the goal and mission of S4AI.

Blockchain, smart contracts, DAO, and DeSci have moved "governance" from the field of liberal arts to the "science and engineering" category of hard technology. New encryption technologies and federation methods, from NFT, Lightning Network, federated learning, federated intelligence to federated ecology, have made

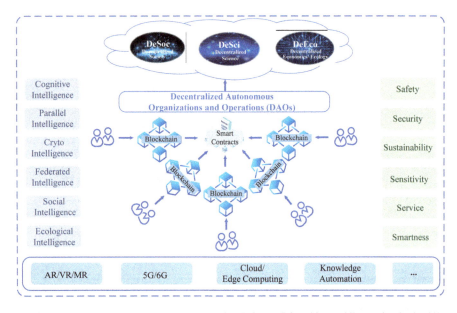

Fig. 9.2 Decentralized ecosystems accompanying Industry 5.0 and its enabling technologies [1]

the governance of intelligent technology a reality, but these technologies are still not enough.

The introduction of "digital scientists" provides a new perspective for the governance of AI4S, that is, the education and cultivation of digital people for scientific research. As envisaged in parallel education research, through digital schools and digital institutes, biological humans and digital scientists can learn and train at the same time in various large educational and scientific models and promote each other to achieve "alignment" and "governance," just like the education and scientific research process experienced by human beings themselves.

"I believe that AI will become a kind of meta-solution for scientists to deploy, enhancing our daily lives and allowing us all to work more quickly and effectively. If we can deploy these tools broadly and fairly, fostering an environment in which everyone can participate in and benefit from them, we have the opportunity to enrich and advance humanity as a whole." said Demis Hassabis, the co-founder and CEO of DeepMind.

But how to achieve the beautiful vision of everyone participating and benefiting? As shown in Fig. 9.2, enabling technologies of the DeSci movement, DAO, and federated ecosystem are paving the way to the goal.

9.2.2 DeSci

Decentralized Science (DeSci) is receiving more and more attention from scientists in different fields [2]. DeSci can play a significant role in building a robust ecosystem for AI4S, offering several potential benefits. While traditional centralized approaches to scientific research have been prevalent, decentralized models bring new possibilities for collaboration, transparency, and inclusivity [3].

DeSci facilitates the sharing of data across decentralized networks, which is crucial in AI4S for training models that are robust, generalizable, and reflective of various scientific scenarios. Additionally, DeSci encourages open access to models, algorithms, and research findings. While federated learning and Swarm Learning can address issues like data privacy and model proprietary rights to some extent, advocating for openness remains essential. This openness in AI4S fosters the development of a shared knowledge base, enabling researchers to build upon each other's work and accelerate advancements in the field.

DeSci promotes collaborative research networks by enabling researchers from diverse geographical locations and institutions to work together, enhancing the diversity of perspectives, expertise, and datasets available for AI4S applications. This collaborative approach fosters community-driven innovation, allowing a broader range of contributors to participate in developing and improving AI4S methodologies, leading to creative solutions and diverse applications. Additionally, decentralization promotes global accessibility to scientific resources and findings in AI4S, democratizing knowledge and technology and enabling researchers from around the world to participate and benefit.

In addition, DeSci offers several key promotions in boosting AI4S:

Increased Transparency and Trust. Decentralization can enhance transparency in the development and deployment of AI models. Transparency is crucial for gaining the trust of the scientific community and ensuring the reproducibility of research findings.

Resilience and Redundancy. Decentralized systems are often more resilient in the face of failures or disruptions. A decentralized ecosystem can provide redundancy in terms of data sources, algorithms, and computational resources, reducing the risk of single points of failure.

Tokenization and Incentives. DeSci can leverage tokenization and incentive mechanisms to reward contributors for sharing data, algorithms, or computational resources. In AI4S, this can encourage collaboration and the development of high-quality, shared resources.

Finally, DeSci allows for adaptability to the specific needs and structures of different scientific domains. This flexibility accommodates the diverse requirements of physics, biology, chemistry, and other disciplines.

While DeSci offers notable advantages, it is essential to consider challenges such as governance, standardization, and coordination. A thoughtful integration of decentralized principles with domain-specific requirements can contribute to the development of a dynamic and collaborative ecosystem for AI4S.

9.2.3 DeSci and DAO

Decentralized Autonomous Organizations (DAOs) and DeSci (Decentralized Science) are both examples of decentralized, community-driven systems. While DeSci specifically focuses on the application of decentralized principles to scientific research and innovation, DAOs are a broader concept that can be applied to various decentralized organizational structures [4, 5].

In the context of DeSci, DAOs could potentially be utilized to govern and manage the decentralized research efforts and collaborations facilitated by the DeSci platform. DAOs could help allocate resources, make decisions about research directions, and govern the overall operation of the platform in a transparent and decentralized manner.

Additionally, DeSci's emphasis on open access, collaboration, and community-driven innovation aligns with the principles of DAOs, which prioritize decentralization, transparency, and community governance. Both DeSci and DAOs aim to empower individuals and communities to participate in decision-making processes and contribute to meaningful advancements in their respective fields.

There are an increasing number of DAOs in different fields, e.g., VitaDAO, ValleyDAO, and AthenaDAO for life and health [6].

9.2.4 DeSci with Blockchain

Beneath the DeSci and DAO are a series of technologies to support the operations. Technologies such as blockchain, distributed ledger technology (DLT), smart contracts, decentralized storage, peer-to-peer networks, and open-access platforms play key roles to enable secure and transparent recording of scientific data, transactions, and collaborations. Tokenization is also essential to incentivize and reward participants in the ecosystem. By embracing these technologies, DeSci has the ability to create a decentralized ecosystem for scientific research that is inclusive and efficient and fosters global collaboration.

Blockchain techniques are crucial for supporting the autonomy, transparency, and efficiency of DAOs, enabling them to operate in a decentralized and democratic manner. Firstly, blockchain provides a decentralized and distributed ledger, ensuring that no single entity controls the entire organization, promoting transparency and preventing censorship or manipulation. Smart contracts, which are self-executing contracts with terms directly written into code, automate certain functions within a DAO, ensuring decisions and transactions follow predefined rules without intermediaries. The transparency of blockchain allows stakeholders to view operations and transactions in real-time, fostering trust and preventing fraud. Additionally, blockchain's immutability ensures the integrity of records and provides a reliable audit trail. Furthermore, blockchain enables secure and transparent voting mechanisms within DAOs, allowing participants to use tokens or other methods to vote on proposals, make decisions, and govern operations.

9.3 Foundation Intelligence Based on TRUE DAO

Currently, the development of artificial intelligence has entered the era of large models. How to coordinate data and computing resources, accelerate core algorithm innovation, promote industrial and social applications, and serve the needs of various industries and groups has become an important topic. It is necessary to fully leverage the capabilities of large models while avoiding the environmental and social issues they may bring. Regulars and incentive policies need to be introduced to address these challenges. To address the issues of data and resources in artificial intelligence, the authors [7] proposed a federated ecosystem framework for artificial intelligence research. Driven by data privacy, information security, and resource integration, the federated ecosystem is essentially built on a series of blockchain-based technologies supporting security, consensus, incentives, and contracts.

9.3.1 Federated Ecosystem

A federated ecosystem consists of four submodules, i.e., federated data, federated control, federated management, and federated services.

(1) Federated data [8] is a key component of the federated ecosystem, addressing the challenge of data silos in the era of large models. It encompasses all nodes' data, storage, computing, and communication resources within the federation. To ensure privacy, federated data is categorized as private or non-private, and data federation is achieved through federated control. In AI applications, federated data supports effective data retrieval, preprocessing, processing, mining, and visualization. It addresses issues like data loss, low quality, and copyright protection in large model training, while ensuring privacy, enabling data sharing, and providing security for public models.

(2) Federated control [9, 10] is the core execution component of the federated ecosystem, ensuring information security and protecting data rights. It employs a distributed strategy for efficient, secure, and reliable control of large systems. Private data remains at local nodes, while ownership and usage rights of non-private data are separated, ensuring security and rights protection. Federated control uses federated contracts to define data federation, establishing control over storage, transmission, sharing, and use of data. Its goal is to ensure information security, break data silos, and achieve data federation, making it crucial for federated intelligence and the overall success of the federated ecosystem.

(3) Federated management [11, 12] is a core part of the federated ecosystem, responsible for making management decisions based on overall goals and adjusting them dynamically in real-time. It helps achieve optimal status and goals of the ecosystem, ensuring security. Federated management provides personalized services and security through control and management of federated data. It utilizes blockchain-based contracts, incentives, and consensus to ensure security while

transforming data into intelligent security. Federated management combines data, computing power, and human resources to make scientifically reliable decisions, improving management efficiency. Supported by AI and blockchain, federated data is aggregated, transformed into decisions, and implemented into measures, achieving the evolution from data to intelligence. Federated intelligence transforms individual intelligence into collective intelligence, and federated management aids in this transformation.

(4) Federated services [1]. The goal of federated management is to achieve federated services through federated control of federated data. Therefore, federated data is the data foundation of federated management and the data security for federated services. By designing a series of federated management rules and ensuring the security and privacy of federated node data, federated services are achieved through the management and control of federated data. At the same time, in the process of achieving federated services, a large amount of new data is constantly generated, which can be added to federated data for optimizing federated management decisions.

9.3.2 *True DAO to Intelligent Federation*

In summary, the federated ecosystem is proposed based on the research idea of the intelligent ecosystem, which has the ability to transform data into intelligence. It is applicable not only to federations dominated by central nodes but also to federations where central nodes are weakened or absent. Through the federated ecosystem, federated nodes can establish cooperative relationships through loose alliances, strengthen the privacy protection of each node, mobilize the enthusiasm of federated nodes, and increase the participation of federated members, thereby improving the overall performance of the federation. The combination of blockchain ("True") and distributed autonomous management ("DAO") forms TAO (True DAO), creating a complete ecosystem of trusted data, algorithms, and operations, providing a guarantee for the development from the federated ecosystem to the intelligent federation.

The framework and methods of the federated ecosystem have been successfully applied in industrial control, transportation logistics, social population, and other fields [13]. Federated ecosystem will also have great potential in AI for science research.

References

1. Wang, X., Yang, J., Wang, Y., et al. (2023). Steps toward industry 5.0: Building "6S" parallel industries with cyber-physical-social intelligence. *IEEE/CAA Journal of Automatica Sinica, 10*(8), 1692–1703.
2. Hamburg, S., et al. (2021). Call to join the decentralized science movement. *Nature, 600*(7888), 221.
3. Ding, W., Hou, J., Li, J., et al. (2022). DeSci based on web3 and DAO: A comprehensive overview and reference model. *IEEE Transactions on Computational Social Systems, 9*(5), 1563–1573.
4. Wang, F.-Y., Ding, W., Wang, X., Garibaldi, J., Teng, S., Imre, R., & Olaverri-Monreal, C. (2022). The DAO to DeSci: AI for free, fair, and responsibility sensitive sciences. *IEEE Intelligent Systems, 37*(2), 16–22.
5. Miao, Q., Zheng, W., Lv, Y., Huang, M., Ding, W., & Wang, F.-Y. (2023). DAO to HANOI via DeSci: AI paradigm shifts from AlphaGo to ChatGPT. *IEEE/CAA Journal of Automatica Sinica, 10*(4), 877–897.
6. (2021). The Community of the DAO. *Nature Biotechnology, 41*(10), 1357.
7. Wang, F.-Y., Qin, R., Chen, Y., et al. (2021). Federated ecology: Steps toward confederated intelligence. *IEEE Transactions on Computational Social Systems, 8*(2), 271–278.
8. Wang, F.-Y., Zhang, W., Tian, Y., et al. (2021). Federated data: Toward new generation of credible and trustable artificial intelligence. *IEEE Transactions on Computational Social Systems, 8*(3), 538–545.
9. Wang, F.-Y., Zhu, J., Qin, R., et al. (2021). Federated control: Toward information security and rights protection. *IEEE Transactions on Computational Social Systems, 8*(4), 793–798.
10. Wang, F.-Y. (2022). The DAO to MetaControl for MetaSystems in metaverses: The system of parallel control systems for knowledge automation and control intelligence in CPSS. *IEEE/CAA Journal of Automatica Sinica, 9*(11), 1899–1908.
11. Wang, F.-Y., Qin, R., Li, J., et al. (2021). Federated management: Toward federated services and federated security in federated ecology. *IEEE Transactions on Computational Social Systems, 8*(6), 1283–1290.
12. Li, J., Qin, R., Wang, F.-Y. (2023). The future of management: DAO to smart organizations and intelligent operations. *IEEE Transactions on Systems, Man, and Cybernetics: Systems, 53*(6), 3389–3399.
13. Wang, F.-Y., Qin, R., Wang, X., et al. (2022). MetaSocieties in metaverse: MetaEconomics and MetaManagement for MetaEnterprises and MetaCities. *IEEE Transactions on Computational Social Systems, 9*(1), 2–7.

Printed in the USA
CPSIA information can be obtained
at www.ICGtesting.com
CBHW072127150924
14543CB00004B/108